Electronic and Optical Properties of Graphite-Related Systems

Electronic and Optical Properties of Graphite-Related Systems

By
Chiun-Yan Lin
Department of Physics, National Cheng Kung University, Taiwan

Rong-Bin Chen
Center of General Studies, National Kaohsiung Marine University, Taiwan

Yen-Hung Ho
Department of Physics, National Tsing Hua University, Taiwan

Ming-Fa Lin
Department of Physics, National Cheng Kung University, Taiwan

CRC Press
Taylor & Francis Group
Boca Raton London New York

CRC Press is an imprint of the
Taylor & Francis Group, an **informa** business

CRC Press
Taylor & Francis Group
6000 Broken Sound Parkway NW, Suite 300
Boca Raton, FL 33487-2742

First issued in paperback 2020

© 2018 by Taylor & Francis Group, LLC

CRC Press is an imprint of Taylor & Francis Group, an Informa business

No claim to original U.S. Government works

ISBN-13: 978-1-138-57106-8 (hbk)
ISBN-13: 978-0-367-65749-9 (pbk)

Visit the Taylor & Francis Web site at
http://www.taylorandfrancis.com

and the CRC Press Web site at
http://www.crcpress.com

Contents

Preface..vii
Authors...ix

1 Introduction 1

2 Theoretical Models and Experimental Techniques 9
  2.1   Magnetic Tight-Binding Model for Layered Graphites.................9
         2.1.1   Simple Hexagonal Graphite...12
         2.1.2   Bernal Graphite 13
         2.1.3   Rhombohedral Graphite 15
         2.1.4   Gradient Approximation for Optical Properties...............18
  2.2   Scanning Tunneling Microscopy and Spectroscopy 22
  2.3   Angle-Resolved Photoemission Spectroscopy..............................24
  2.4   Absorption Spectroscopy..25

3 Simple Hexagonal Graphite 29
  3.1   Electronic Structures without External Fields..............................29
  3.2   Optical Properties without External Fields...................................34
  3.3   Magnetic Quantization 36
         3.3.1   Landau Levels and Wave Functions 36
         3.3.2   Landau-Subband Energy Spectra..38
  3.4   Magneto-Optical Properties...43

4 Bernal Graphite..51
  4.1   Electronic Structures without External Fields..............................51
  4.2   Optical Properties without External Fields...................................54
  4.3   Magnetic Quantization 57
         4.3.1   Landau Subbands and Wave Functions 57
         4.3.2   Anti-Crossings of Landau Subbands..................................60
  4.4   Magneto-Optical Properties...62

5 Rhombohedral Graphite 69
  5.1   Electronic Structures without External Fields..............................69
  5.2   Anisotropic Dirac Cone Along a Nodal Spiral..............................71
  5.3   Dimensional Crossover...73
  5.4   Optical Properties without External Fields...................................75
  5.5   Magneto-Electronic Properties 76
         5.5.1   Tight-Binding Model 76
         5.5.2   Onsager Quantization...78
  5.6   Magneto-Optical Properties...82

6 **Quantum Confinement in Carbon Nanotubes and Graphene Nanoribbons** 85
   6.1   Magneto-Electronic Properties of Carbon Nanotubes 85
   6.2   Magneto-Optical Spectra of Carbon Nanotubes............................91
   6.3   Magneto-Electronic Properties of Graphene Nanoribbons 96
   6.4   Magneto-Optical Spectra of Graphene Nanoribbons....................99
   6.5   Comparisons and Applications 107

7 **Concluding Remarks** 113

Problems 121

References 123

Index 147

Preface

This book is intended as an introduction to the electronic and optical properties of graphite-related systems. A systematic review is made for AA-, AB- and ABC-stacked graphites. The generalized tight-binding model, accompanied by the effective-mass approximation and the Kubo formula, is developed to investigate electronic and optical properties in the presence/absence of a uniform magnetic field. The unusual electronic properties cover the stacking-dependent Dirac-cone structures, the significant energy widths along the stacking direction, the Landau subbands crossing the Fermi level, the magnetic-field-dependent Landau-subband energy spectra with crossings and anti-crossings, and the monolayer- or bilayer-like Landau wave functions. There are configuration-created special structures in density of states and optical spectra. The three kinds of graphites are quite different from one another in terms of the available inter-Landau-subband excitation channels, including the number, frequency, intensity and structure of absorption peaks. The dimensional crossover presents the main similarities and differences between graphites and graphenes; furthermore, the quantum confinement enriches the magnetic quantization phenomena in carbon nanotubes and graphene nanoribbons. The cooperative/competitive relations between the interlayer atomic interactions, dimensions and magnetic quantization are responsible for the diversified essential properties. Some of the theoretical predictions are consistent with the experimental measurements.

Chiun-Yan Lin, Rong-Bin Chen,
Yen-Hung Ho, and Ming-Fa Lin

Authors

 **Rong-Bin Chen** is a professor at the Center of General Studies, National Kaohsiung Marine University, Taiwan. He received his PhD degree in physics from the University of Cheng Kung in 2000 studying III-V semiconductor physics. His main research interests are condensed matter theory; electronic and optical properties of graphene, low-dimensional materials, and carbon-related systems.

 **Yen-Hung Ho** received his PhD in physics in 2009 from the National Cheng-Kung University, Taiwan. Currently, he is a postdoctoral researcher at the National Center for Theoretical Sciences and Physics Department in National Tsing-Hua University. His research mainly focuses on the model calculations of electronic properties for nanomaterials in response to external magnetic, electric and strain fields.

 **Chiun-Yan Lin** obtained his PhD in 2014 in physics from the National Cheng Kung University (NCKU), Taiwan. Since 2014, he has been a postdoctoral researcher in the department of physics at NCKU. His main scientific interests are in the field of condensed matter physics, modeling and simulation of nanomaterials. Most of his research is focused on the electronic and optical properties of two dimensional nanomaterials.

 **Ming-Fa Lin** is a distinguished professor in the Department of Physics, National Cheng Kung University, Taiwan. He received his PhD in physics in 1993 from the National Tsing-Hua University, Taiwan. His main scientific interests focus on essential properties of carbon related materials and low-dimensional systems.

1

Introduction

Carbon atoms can form various condensed-matter systems with unique geometric structures, mainly owing to four active atomic orbitals. Zero-to three-dimensional carbon-related systems cover diamond [1], graphite [2–4], graphene [5], graphene nanoribbons [6], carbon nanotubes [7] and carbon fullerene [8]. All of these have sp^2 bonding except for diamond which has sp^3. The former might exhibit similar physical properties, for example, π-electronic optical excitations [9,10]. Graphite is one of the most extensively studied materials, theoretically and experimentally. This layered system is very suitable for exploring diverse 3D and 2D phenomena. The interplane attractive forces originate from the weak Van der Waals interactions of the $2p_z$ orbitals. The honeycomb lattice and the stacking configuration are responsible for the unique properties of graphite, for example, its semimetallic behavior due to the hexagonal symmetry and the interlayer atomic interactions. The essential properties are dramatically changed by the intercalation of various atoms and molecules. Graphite intercalation compounds can achieve a conductivity as good as copper [4,11–13]. In general, there exist three kinds of ordered configurations in the layered graphites and compounds, namely AA, AB and ABC stackings. Simple hexagonal (AA), Bernal (AB) and rhombohedral graphites (ABC) exhibit rich and diverse electronic and optical properties in the presence/absence of a uniform magnetic field ($\mathbf{B} = B_0\hat{z}$). To present a systematic review of them, the generalized tight-binding model is developed under magnetic quantization. This model, combined with the Kubo formula, is utilized to investigate the essential properties of layered carbon-related systems. The dimensional crossover from graphene to graphite and the quantum confinement in nanotube and nanoribbon systems are discussed thoroughly. A detailed comparison with other theoretical studies and experimental measurements is also made.

Few-layer and multilayer graphenes, with distinct stacking configurations, have been successfully produced using various experimental methods since the first discovery of monolayer graphene in 2004 by mechanical exfoliation [5]. They possess a hexagonal symmetry and a nanoscale size, leading to many remarkable characteristics, for example, they have the largest Young's modulus [15], feature-rich energy bands [16,17], diverse optical selective rules [18–25], unique magnetic quantization [27–36], anomalous quantum Hall effects (QHE) [37–41] and multimode plasmons [42–45]. Their electronic and optical properties are very sensitive to changes to the stacking configuration [16,29,31,33,34,36], layer number [16,29,31,36], magnetic field [17–24,27–32,36], electric field [17,32], mechanical strain [16,17], doping [16,17] and sliding [35]. Five kinds of electronic structures—linear [36], parabolic [27,31,36], partially flat [31,36], sombrero-shaped [31,32,36] and oscillatory [32,34,35] energy bands—are revealed in AB- and ABC-stacking systems. Oscillatory bands can be created by a perpendicular electric field. However, the AA stacking only has the first kind of electronic structure. Specifically, the intersection of the linear valence and conduction bands can form the so-called Dirac-cone structure. The main features of energy bands are directly reflected in the other essential properties. The finite-layer confinement effects are expected to induce important differences between the layered graphenes and graphites. The close relations arising from the dimensional crossover deserve a thorough investigation.

Graphite crystals are made up of a series of stacked graphene planes. Among three kinds of ordered stacking configurations, AB-stacked graphite is predicted to have the lowest ground state energy according to the first-principles calculations [46]. Natural graphite presents dominant AB stacking and partial ABC stacking [2,3]. AA-stacked graphite, which possesses the simplest crystal structure, does not exist naturally. The periodical AA stacking is first observed in the Li-intercalation graphite compounds [4] with a high free electron density and a super conducting transition temperature of 1.9 K [14]. Simple hexagonal graphite is successfully synthesized by using DC plasma in hydrogen-methane mixtures [47], and AA-stacked graphenes are generated using the method of Hummers and Offeman and chemical-vapor deposition (CVD) [48,49]. Furthermore, the AA-stacking sequence is confirmed through high-resolution transmission electron microscopy (HRTEM) [48]. Specifically, angle-resolved photoemission spectroscopy (ARPES), a powerful tool in the direct identification of energy bands, is utilized to examine two/three pairs of Dirac-cone structures in AA-stacked bilayers/trilayers [50,51].

AA-stacking systems have stirred many theoretical research studies, such as the investigation of band structures [29,31,52,53,53], magnetic quantization [29,31,36,54,55], optical properties [56–58], Coulomb excitations [59,61–63], quantum transport [66] and phonon spectra [67]. Simple hexagonal graphite, a 3D-layered system with the same graphitic sheets on the (x, y)-plane, was first proposed by Charlier et al. [46]. From the tight-binding model and the first-principles method, this system belongs to a band-overlap semimetal, in which the same electron and hole density originate from the significant interlayer atomic interactions. There is one pair of low-lying valence and conduction bands. The critical feature is the vertical Dirac-cone structure with a sufficiently wide bandwidth of ~ 1 eV along the k_z-direction (**k** wave vector). Similarly, the AA-stacked graphene is predicted to exhibit multiple Dirac cones vertical to one another [29,36,52]. Moreover, Bloch wave functions are the only symmetric or anti-symmetric superpositions of the tight-binding functions on the distinct sublattices and layers [36]. Apparently, these 3D and 2D vertical Dirac cones will dominate the other low-energy essential properties, for example, the special structures in density of states (DOS) [36,53,54], the quantized Landau subbands (LSs) and levels (LLs) [29,36,55], the rich optical spectra [57,58] and the diversified plasmon modes [59,61–63]. The magnetic quantization is frequently explored by the effective-mass approximation [29] and the generalized tight-binding model [31,36]. It is initiated from the vertical Dirac points: the extreme points in the energy-wave-vector space. This creates the specific B_0-dependent energy spectra and the well-behaved charge distributions, thus leading to the diverse and unique magneto-absorption peaks, for example, the intraband and interband inter-LS excitations, the multichannel threshold peaks and the beating-form absorption peaks in AA-stacked graphite [55,57,58]. The predicted band structures, energy spectra and optical excitations can be verified by ARPES [50,51], scanning tunneling spectroscopy (STS) [64,65] and optical spectroscopy [20–24], respectively.

Bernal graphite is a well-known semimetal [2] with a conduction electron concentration of ~5×10^{18}/cm^3. Furthermore, the AB-stacking configuration is frequently observed in the layered systems, for example, bilayer [5,68], trilayer [68–70] and tetralayer graphenes [68]. AB-stacked graphite possesses two pairs of low-lying energy bands that are dominated by the $2p_z$ orbitals, owing to a primitive unit cell with two neighboring layers. The highly anisotropic band structure, the strong and weak energy dispersions along the (k_x, k_y) plane and k_z-axis, respectively, is confirmed by the ARPES measurements [71–75]. Similar examinations are done for two pairs of parabolic bands in a bilayer AB stacking [78,79], and the linear and parabolic bands in a trilayer system [78,80]. The measured DOS of Bernal graphite presents the splitting π and π^* peaks at the middle energy (approximately 0.5eV–0.8eV) [64], reflecting the highly accumulated states near the saddle points. Furthermore, it is finite near the Fermi level (E_f) because of the semimetallic behavior [65]. An electric-field-induced band gap is observed in the bilayer AB stacking [69]. The magnetically quantized energy spectra, with many special structures, are also identified using the STS measurements for AB-stacked graphite [81,82] and graphenes [83,84], especially the square-root and linear dependences on the magnetic field strength (the monolayer- and bilayer-like behaviors at low energy). From the optical measurements, Bernal graphite shows a very prominent π-electronic absorption peak at frequency ~5 eV [85], as revealed in carbon-related systems with the sp^2 bonding [86,87]. Concerning the low-frequency magneto-optical experiments, the measured excitation spectra due to LSs [88–92,94,96] or LLs [18–24] clearly reveal a lot of pronounced absorption structures, the selection rule of $\Delta n = \pm 1$ (n quantum number) and the strong dependence on the wave vector of k_z or the layer number (N).

The earliest attempt to calculate the band structures of monolayer graphene and Bernal graphite was made by Wallace using the tight-binding model with the atomic interactions of $2p_z$ orbitals [97]. The former 2D system has the linear valence and conduction bands intersecting at E_F, where it belongs to a zero-gap semiconductor with a vanishing DOS. However, the semimetallic, 3D electronic structure can be further comprehended through the Slonczewski–Weiss–McClure Hamiltonian involving important intra-layer and interlayer atomic interactions [98,99]. The magnetic Hamiltonian can be solved by the low-energy perturbation approximation, in which the LS energy spectra exhibit crossing and anti-crossing behaviors [101,102]. The generalized tight-binding model, which deals with the magnetic field and all atomic interactions simultaneously, was developed to explore the main features of LSs, for example, two groups of valence and conduction LSs, and the layer-, k_z- and B_0-dependent spatial oscillation modes [88,100]. As to the AB-stacked graphenes, their electronic and magneto-electronic properties present bilayer- and monolayer-like behaviors, which are associated with two pairs of parabolic bands and a slightly distorted Dirac cone, respectively [16,36]. The former is only revealed in the odd-N systems. Optical spectra of AB-stacked systems are predicted to exhibit a strong dependence of special structures on the layer number and dimension [18–24]. The magneto-optical excitations arising from two groups of LSs in graphite or N groups of LLs in graphenes could be evaluated using the generalized tight-binding model [88]. On the other hand, those due to the first group of LSs/LLs are frequently investigated in detail by the effective-mass approximation [101,102]. The calculated electronic and optical properties are in agreement with the experimental measurements [81,82].

The rhombohedral phase is usually found to be mixed with the Bernal phase in natural graphite. The ABC-stacking sequence in bulk graphite is directly identified from the experimental measurements of HRTEM [103], x-ray diffraction [104–106] and scanning tunneling microscopy (STM) [107,108]. Specifically, this stacking configuration can account for the measured 3D QHE with multiple plateau structures [109] as rhombohedral graphite possesses well-separated LS energy spectra [111,112]. In addition, doped Bernal graphite is predicted to exhibit only one plateau [113]. Both ABC- and AB-stacked graphenes can be produced by the mechanical exfoliation of kish graphite [114,115], CVD [116–119], chemical and electrochemical reduction of graphite oxide [120–122], arc discharge [123–125], flame synthesis [126] and electrostatic manipulation of STM [127,130]. The ARPES measurements have verified the partially flat, sombrero-shaped and linear bands in trilayer ABC-stacking [80]. As to the STS spectra, a pronounced peak at the Fermi level characteristic of the partial flat band is revealed in trilayer and penta-layer ABC stacking [128–130]. Moreover, infrared reflection spectroscopy and absorption spectroscopy are utilized to examine the low-frequency optical properties, displaying clear evidence of two-featured absorption structures due to the partially flat and sombrero-shaped energy bands [131]. According to the specific infrared conductivities, infrared scattering scanning near-field optical microscopy can distinguish the ABC-stacking domains with nanoscaled resolution from other domains. However, the magneto-optical measurements for ABC-stacked graphenes have been absent up to now.

ABC-stacked graphite has a rhombohedral unit cell, while AA- and AB-stacked systems possess hexagonal ones. This critical difference in stacking symmetry is responsible for the diversified essential properties. Among three kinds of bulk graphites, rhombohedral graphite is expected to present the smallest band overlap (the lowest free carrier density), and the weakest energy dispersion along the k_z-direction [110,132]. There also exists a robust low-energy electronic structure: a 3D spiral Dirac-cone structure. This results in unusual magnetic quantization [111,112,134], in which the quantized LSs exhibit monolayer-like behavior and significant k_z-dependence. Previous studies show that four kinds of energy dispersions exist in ABC-stacked graphenes [31,36]. Specifically, the partially flat bands corresponding to the surface states and the sombrero-shaped bands are absent in the bulk system. They can create diverse and unique LLs with asymmetric energy spectra around E_F, normal and abnormal B_0-dependences, well-behaved and distorted probability distributions, and frequent crossings and anti-crossings [31,36]. Apparently, optical and magneto-optical properties are greatly enriched by the layer number and dimension [135–138]. The quantized LLs of the partially flat bands and the lowest sombrero-shaped band have been verified by magneto-Raman spectroscopy for a large ABC domain in a graphene multilayer flake [139]. Layered graphenes are predicted to have more complicated excitation spectra compared with 3D systems. The former and the latter, respectively, reveal N^2 categories of inter-LL transitions and one category of inter-LS excitations [136,137].

In addition to the stacking configurations, the distinct dimensions can create diverse phenomena in carbon-related systems. Quantum confinement in 1D carbon nanotubes and graphene nanoribbons plays a critical role in the essential properties. Systematic studies have been made for the former since the successful synthesis of carbon nanotubes using arc-discharge evaporation in 1991 [7]. Each carbon nanotube can be regarded as a rolled-up graphene sheet in a cylindrical form. It is identified as a metal or semiconductor, depending on the radius and chiral angle [140–142]. The geometry-dependent energy spectra, with energy gaps (E_g's), are directly verified from the STS measurements [207,208]. Specifically, the cylindrical symmetry can present the well-known Aharonov–Bohm effect under an axial magnetic field [145–148]. This is confirmed by the experimental measurements on optical [149,150] and transport properties [151–153]. However, a closed surface acts as a high barrier in the formation of the dispersionless LLs, as a perpendicular magnetic field leads to a vanishing flux through carbon hexagons. It is very difficult to observe the physical phenomena associated with the highly degenerate states, except for very high field strength [154].

The essential properties are greatly enriched by the boundary conditions in 1D systems. The open and periodical boundaries, which, respectively, correspond to the graphene nanoribbon and carbon nanotube, induce the important differences between them. A graphene nanoribbon is a finite-width graphene or an unzipped carbon nanotube. Graphene nanoribbons can be produced by cutting few-layer graphenes [155,156], unzipping multiwalled carbon nanotubes [157–159] and using direct chemical syntheses [160–162]. The cooperative or competitive relations between the open boundary, the edge structure and the magnetic field are responsible for the rich and unique properties. The 1D parabolic bands, with energy gaps, in armchair graphene nanoribbons are confirmed by ARPES [163]. Furthermore, STS has verified the asymmetric DOS peaks and the finite-size effect on the energy gap [164–167]. Optical spectra are predicted to have edge-dependent selection rules [168–170]. The theoretical calculations show that only quasi-LLs (QLLs), with partially dispersionless relations, can survive in the presence of a perpendicular magnetic field [171,172]. The magneto-optical selection rule, as revealed in layered graphenes, sharply contrasts with that of the carbon nanotubes with well-defined angular momenta along the azimuthal direction [173].

In this book, we propose and develop a generalized tight-binding model to fully comprehend the electronic and optical properties of graphite-related systems. The Hamiltonian is built from the tight-binding functions on the distinct sublattices and layers, in which all important atomic interactions, stacking configurations, layer numbers and external fields are taken into account simultaneously. A rather large Hamiltonian matrix, being associated with the periodical variation of the vector potential, is solved using an exact diagonalization method. The essential properties can be evaluated very efficiently. Moreover, the effective-mass approximation is utilized to provide the qualitative behaviors and the semiquantitative results, for example, the layer-dependent characteristics. Specifically, the Onsager quantization method is introduced to study the magnetic LS energy spectra in ABC-stacked graphite with unique spiral Dirac cones. Such approximations are useful in the identification of the critical atomic interactions that create the unusual properties.

The AA-, AB- and ABC-stacked graphites and graphenes, and 1D graphene nanoribbons and carbon nanotubes, are worthy of a systematic review of essential properties. Electronic and optical properties, which mainly come from carbon $2p_z$ orbitals, are investigated in the presence/absence of a magnetic field. Electronic structures, quantized LS and LL state energies, magnetic wave functions, DOS and optical spectral functions are included in the calculated results. Band widths, energy dispersion relations, critical points in energy-wave-vector space, crossings and anti-crossings of B_0-dependent energy spectra, spatial oscillation modes of localized probability distributions, and various special structures in DOS are explored in detail. The main features of optical excitations focus on the available excitation channels; the form, number, intensity and frequency of prominent absorption structures; and the layer/dimension and field dependences. Moreover, the theoretical predictions are compared with the ARPES, STS and optical measurements and require more experimental examinations. Chapter 2 covers geometric structures, important atomic interactions, the generalized tight-binding model and the Kubo formula, in which the main issues are the construction of the magnetic Hamiltonian and the efficient combination of two methods. In Chapter 3, the stacking- and layer-enriched essential properties are studied for the AA-stacked graphite and graphenes, especially those due to multiple vertical Dirac-cone structures. The analytical band structures and magneto-electronic energy spectra are obtained from the approximate expansions around the high symmetry points. They illustrate the diversified characteristics of the close relations between, for example, the determination of the close relations between the absorption spectra and the important intralayer/interlayer atomic interactions. The dimensional crossover from monolayer graphene to graphite creates the critical differences between 2D and 3D phenomena, including the semiconductor–semimetal transition, the k_z-dependent band width, the LS/LL energy spectra near the Fermi level, the optical gap and the low- and middle-frequency absorption structures.

The dramatic transformations of essential properties are clearly revealed in distinct stacking configurations. As to the AB-stacked systems, the linear and parabolic energy dispersions, crossings and anti-crossings of LS/LL energy spectra, well-behaved and perturbed magnetic wave functions, layer- and dimension-dependent optical spectra, and rich magneto-absorption peaks are investigated in Chapter 4. The monolayer- and bilayer-like behaviors are presented for Bernal graphite and layered AB stacking. Specifically, ABC-stacked graphene has linear, parabolic, partially flat and sombrero-shaped energy bands, while rhombohedral graphite exhibits a 3D spiral Dirac-cone structure, as indicated in Chapter 5. Such characteristics are expected to create unique essential properties. The low-energy approximation and magnetic quantization are proposed to explain the diversified electronic properties and optical spectra. In Chapter 6, the reduced dimension in graphene nanoribbons and carbon nanotubes leads to the rich essential properties being sensitive to the open/periodical boundary condition, width/radius, edge/chiral angle and external fields. Comparisons among the graphite-related systems and potential applications are also discussed. Finally, Chapter 7 contains concluding remarks and discusses how the theoretical framework could be further extended to other mainstream layered materials.

2

Theoretical Models and Experimental Techniques

In the presence of a uniform magnetic field, $\mathbf{B} = B_0\hat{z}$, electrons are forced to undergo cyclotron motion in the x–y plane. As a result, electronic states are evolved into highly degenerate states, referred to as Landau subbands (LSs) in graphites. The 1D LSs are calculated from the subenvelope functions established on different sublattices in the framework of the generalized tight-binding model, which simultaneously takes into account external fields and atomic interactions. The magneto-Hamiltonian is built from the tight-binding functions coupled with a periodic Peierls phase in an enlarged unit cell; the period depends on the commensurate relation between the lattice constant and the Peierls phase. According to the Kubo formula, it could be further utilized to comprehend the main features of magneto-absorption spectra, which are closely related to the Landau-level spectrum and the transition matrix elements. The method provides accurate and reliable results for a wide energy range. The three prototypical configurations of bulk graphites, namely simple hexagonal, Bernal and rhombohedral graphites, are chosen for a systematic review. The magnetic quantization in 3D graphene systems shows interesting phenomena as a function of the stacking configuration and the magnetic field strength. The aforementioned electronic and optical properties can be experimentally verified by scanning tunneling microscopy (STM), scanning tunneling spectroscopy (STS), angle-resolved photoemission spectroscopy (ARPES) and absorption spectroscopy. A brief introduction of these instruments will be presented.

2.1 Magnetic Tight-Binding Model for Layered Graphites

The geometric structures of simple hexagonal, Bernal and rhombohedral bulk graphites are shown in Figure 2.1a through c. They are, respectively, constructed from 2D graphene layers periodically stacked along $\hat{z}$ with AA-, AB- and ABC-stacking configurations, where the layer–layer distance I_z is set as 3.35 Å. Detailed definitions of the stacking sequences are provided in the following sections. The hexagonal unit cells of different graphites are marked by the gray shadows, which contain two sublattices, A^l and B^l, on

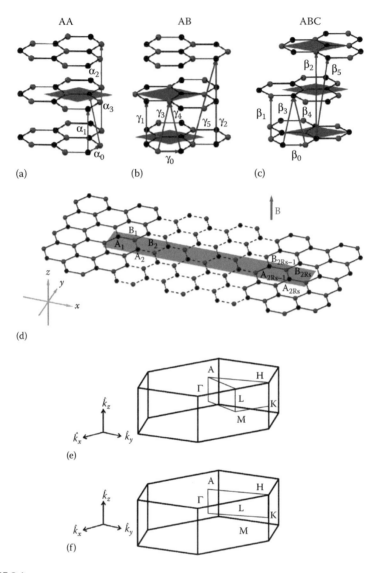

FIGURE 2.1
Geometric structures and atomic interactions of (a) AA-, (b) AB- and (c) ABC-stacked graphites; (d) an enlarged unit cell in a magnetic field; (e) the zero-field first Brillouin zone; and (f) the field-dependent reduced zone.

each layer, where l represents the number of the layer, and the symbols α, β, and γ indicate the interlayer and intralayer hopping integrals. The first Brillouin zone is a hexagonal prism, as shown in Figure 2.1e, where the highly symmetric points are defined as Γ, M, K, A, L and H. The KH lengths for the AA, AB and ABC stackings are, respectively, equal to π/I_z, $\pi/2I_z$ and $\pi/3I_z$ based on the periods along $\hat{z}$.

In general, the essential physical properties are mainly determined by $2p_z$ orbitals of carbon atoms. Built from the subspace spanned by the tight-binding functions φ_{A^l} and φ_{B^l} ($l = 1, 2, \ldots$), the wave function is characterized by their linear combination over all A and B sublattices in a unit cell:

$$\Psi = \sum_l C_{A^l} \varphi_{A^l} + C_{B^l} \varphi_{B^l}, \tag{2.1}$$

where C_{A^l} and C_{B^l} are normalization factors. The tight-binding functions are

$$\varphi_{A^l} = \sum_{R_{A^l}} \exp\left(i\mathbf{k} \cdot \mathbf{R}_{A^l}\right) \chi\left(\mathbf{r} - \mathbf{R}_{A^l}\right)$$
$$\varphi_{B^l} = \sum_{R_{B^l}} \exp\left(i\mathbf{k} \cdot \mathbf{R}_{B^l}\right) \chi\left(\mathbf{r} - \mathbf{R}_{B^l}\right), \tag{2.2}$$

where:
$\chi(\mathbf{r})$ is the atomic $2p_z$ orbital of an isolated carbon
$\mathbf{R}$ is the position vector of an atom

The effective momentum in the presence of $\mathbf{B}$ is $\mathbf{P} - e\mathbf{A}/c$, so that the 3D electronic bands of bulk graphites are quantized into the so-called 1D LSs. A periodic Peierls phase $G_R \equiv 2\pi/\phi_0 \int_{R'}^{R} \mathbf{A}(\mathbf{r}) d\mathbf{r}$ is introduced to the tight-binding functions in Equation 2.2, where $\mathbf{A}$ is the vector potential and $\phi_0 = 2\pi\hbar c/e$ (4.1356×10^{-15} [T·m^2]) is the flux quantum. The Hamiltonian element coupled with the Peierls phase factor is given by

$$H_{i,j}^{B} = H_{i,j} e^{i\Delta G_{i,j}} = H_{i,j} e^{i\frac{2\pi}{\phi_0} \int_{R_j}^{R_j} \mathbf{A}(\mathbf{r}) \cdot d\mathbf{r}}. \tag{2.3}$$

The phase factor gives rise to an enlargement of the primitive unit cell (Figure 2.1d), depending on the commensurate period of the lattice and the Peierls phase. Using the Landau gauge $\mathbf{A} = (0, B_0 x, 0)$, the period of the phase is $l = 3R_B b \hat{x}$, in which there are $2R_B$ A and $2R_B$ B atoms in an enlarged unit cell $\left(R_B = \dfrac{\phi_0 / \left(3\sqrt{3}b'^2/2\right)}{B_0} \simeq \dfrac{79000\ \text{T}}{B_0} \right)$.

This implies that the wave functions of graphites under a uniform magnetic field can be characterized by the subenvelope functions spanned over all bases in the enlarged unit cell, the zero points of which are used to define the quantum numbers of LSs. The wave function is decomposed into two components in the magnetically enlarged unit cell as follows:

$$|\Psi_{\mathbf{k}}\rangle = \sum_{m=1}^{2R_B - 1} \left(A_o |A_{mk}\rangle + B_o |B_{mk}\rangle\right) + \sum_{m=1}^{2R_B} \left(A_e |A_{mk}\rangle + B_e |B_{mk}\rangle\right), \tag{2.4}$$

where o and e, respectively, represent the odd-indexed and even-indexed parts. The subenvelope function $A_{o,e}$ ($B_{o,e}$), described by an n-th order Hermite polynomial multiplied with a Gaussian function, is an even or odd function (depending on n) and represents the probability amplitude of the wave function contributed by each carbon atom. Considering the Peierls substitution for interlayer and intralayer atomic interactions, we can obtain the explicit form of the magnetic Hamiltonian matrix of bulk graphites. A procedure for the band-like Hamiltonian matrix is further introduced to efficiently solve the eigenvectors and eigenvalues by choosing an appropriate sequence for the bases. In the following sections, the Hamiltonian matrices are derived for simple hexagonal, Bernal and rhombohedral graphites in the generalized tight-binding model.

2.1.1 Simple Hexagonal Graphite

In simple hexagonal graphite, each graphene sheet has the same projection on the x–y plane, as shown in Figure 2.1a. The primitive unit cell includes only two atoms that are the same as those of monolayer graphene. Four important atomic interactions are used to describe the electronic properties, that is, $\alpha_0(= 2.569$ eV$)$, $\alpha_1(= 0.361$ eV$)$, $\alpha_2(= 0.013$ eV$)$ and $\alpha_3(=-0.032$ eV$)$ [53], respectively, derived from intralayer hopping between nearest-neighbor atoms, interlayer vertical hoppings between nearest- and next-nearest-neighbor planes, and non-vertical hopping between nearest-neighbor planes. The zero-field Hamiltonian matrix in the subspace of a tight-binding basis $\{\varphi_A, \varphi_B\}$ is expressed as

$$H_{AA} = \begin{pmatrix} \alpha_1 h + \alpha_2 \left(h^2 - 2 \right) & \left(\alpha_0 + \alpha_3 h \right) f \left(k_x, k_y \right) \\ \left(\alpha_0 + \alpha_3 h \right) f^* \left(k_x, k_y \right) & \alpha_1 h + \alpha_2 \left(h^2 - 2 \right) \end{pmatrix}, \quad (2.5)$$

where:

$$f(k_x,k_y) = \sum_{j=1}^{3} \exp(i\mathbf{k} \cdot \mathbf{r}_j) = \exp(ibk_x) + \exp(ibk_x/2)\cos(\sqrt{3}bk_y/2) \quad \text{represents}$$

the phase summation arising from the three nearest neighbors
$h = 2\cos(k_z I_z)$

The k_z-dependent terms are involved in the matrix elements due to the periodicity along the z-direction. The π-electronic energy dispersions are obtained from diagonalizing the Hamiltonian matrix in Equation 2.5:

$$E_{\pm}^{c,v} \left(k_x, k_y, k_z \right) = \alpha_1 h + 2\alpha_2 \left(\frac{h^2}{2-1} \right) \pm \left(\alpha_0 + \alpha_3 h \right) \left| f \left(k_x, k_y \right) \right|, \quad (2.6)$$

and wave functions are

$$\Psi_{\pm,k}^{c,v} = \frac{1}{\sqrt{2}}\left[\varphi_{A^l,k} \pm \frac{f^*(k_x,k_y)}{|f(k_x,k_y)|}\varphi_{A^l,k}\right]. \tag{2.7}$$

The superscripts c and v, respectively, represent the conduction and valence states.

As a result of the vector-potential-induced phase, the number of bases in the primitive unit cell is increased by $2R_B$ times compared to the zero-field case. Using the Peierls substitution of Equation 2.4 and considering only the neighboring atoms coupled by α's, one can derive a band-like form for the magnetic Hamiltonian matrix of AA-stacked graphite:

$$\langle B_{mk} | H | B_{m'k}\rangle = \langle A_{mk} | H | A_{m'k}\rangle = \left[\alpha_1 h + \alpha_2\left(h^2 - 2\right)\right]\delta_{m,m'}, \tag{2.8}$$

$$\langle A_{mk} | H | B_{m'k}\rangle = (\alpha_0 + \alpha_3 h)\left(t_{1k}(m)\delta_{m,m'} + q\delta_{m-1,m'}\right). \tag{2.9}$$

The eigenvector is expanded in the bases with the specific sequence $\{A_{1k}, B_{2R_Bk}, B_{1k}, A_{2R_Bk}, \ldots\ldots B_{R_Bk}, A_{R_B+1k}\}$ and the three independent phase terms are expressed as

$$t_{1k}(m) = \exp\left\{i\left[-(k_x b/2)-\left(\sqrt{3}k_y b/2\right)+\pi\Phi(m-1+1/6)\right]\right\}+$$

$$\exp\left\{i\left[-(k_x b/2)+\left(\sqrt{3}k_y b/2\right)-\pi\Phi(m-1+1/6)\right]\right\}, \tag{2.10}$$

$$q = \exp(ik_x b),$$

where Φ is the magnetic flux through each hexagon. By diagonalizing the Hamiltonian matrix, the k_z-dependent energies and wave functions of the valence and conduction LSs are thus obtained. Such a band-like matrix spanned by a specific order of the bases is also applicable to other prototype bulk graphites. In addition, when the low-energy approximation, in relation to the Dirac points, is made for the Hamiltonian matrix in Equation 2.5, the LS's spectrum is further evaluated from magnetic quantization (discussion in Section 3.3.2) [55]. However, the conservation of 3D carrier density needs to be included in this evaluation.

2.1.2 Bernal Graphite

Bernal graphite is the primary component of natural graphite. The primitive unit cell comprises A^1, B^1, A^2 and B^2 atoms on two adjacent layers, where A^1 and A^2 (B^1 and B^2) are directly located above or below A^2 and A^1 (the centers of hexagons) in adjacent layers, a configuration named AB stacking (Figure 2.1b). The critical atomic interactions based on the

Slonczewski–Weiss–McClure (SWM) model cover γ_0, , γ_5, which are interpreted as hopping integrals between nearest-neighbor and next-nearest-neighbor atoms, and additionally γ_6, which refers to the difference of the chemical environments between non-equivalent A and B atoms. The values are as follows: $\gamma_0 = 3.12$ eV, $\gamma_1 = 0.38$ eV, $\gamma_2 = -0.021$ eV, $\gamma_3 = 0.28$ eV, $\gamma_4 = 0.12$ eV, $\gamma_5 = -0.003$ eV and $\gamma_6 = -0.0366$ eV [46].

The tight-binding Hamiltonian is described by a 4×4 matrix, which, expanded in the basis $\{\varphi_A^1, \varphi_B^1, \varphi_B^2, \varphi_A^2\}$, takes the form

$$H_{AB} = \begin{pmatrix} E_A & \gamma_0 f(k_x,k_y) & \gamma_1 h & \gamma_4 h f^*(k_x,k_y) \\ \gamma_0 f^*(k_x,k_y) & E_B & \gamma_4 h f^*(k_x,k_y) & \gamma_3 h f(k_x,k_y) \\ \gamma_1 h & \gamma_4 h f(k_x,k_y) & E_A & \gamma_0 f^*(k_x,k_y) \\ \gamma_4 h f(k_x,k_y) & \gamma_3 h f^*(k_x,k_y) & \gamma_0 f(k_x,k_y) & E_B \end{pmatrix} \% \tag{2.11}$$

where $E_A = \gamma_6 + \gamma_5 h^2/2$ and $E_B = \gamma_2 h^2/2$. These indicate the sum of the on-site energy and hopping energy of A and B atoms, respectively.

Energy bands and wave functions are easily calculated by diagonalizing the Hamiltonian matrix.

At $\mathbf{B} = B_0 \hat{z}$, the magnetically enlarged unit cell includes $2 \times 4R_B$ bases, which constructs the k_z-dependent Hamiltonian matrix with non-zero terms only between neighboring sublattices on same and different layers. An explicit form of the matrix element is given by

$$\left\langle B_{mk}^1 \mid H \mid A_{m'k}^1 \right\rangle = -\gamma_0 \left(t_{1k}(m)\delta_{m,m'} + q\delta_{m+1,m'} \right), \tag{2.12}$$

$$\left\langle B_{mk}^1 \mid H \mid A_{m'k}^2 \right\rangle = \gamma_4 h \left(t_{1k}(m)\delta_{m,m'} + q\delta_{m+1,m'} \right), \tag{2.13}$$

$$\left\langle A_{mk}^2 \mid H \mid A_{m'k}^1 \right\rangle = \gamma_1 h \delta_{m,m'}, \tag{2.14}$$

$$\left\langle B_{mk}^2 \mid H \mid B_{m'k}^1 \right\rangle = \gamma_3 h \left(t_{2k}(m)\delta_{m,m'} + q\delta_{m+1,m'} \right), \tag{2.15}$$

$$\left\langle A_{mk}^2 \mid H \mid B_{m'k}^2 \right\rangle = -\gamma_0 \left(t_{3k}(m)\delta_{m-1,m'} + q\delta_{m,m'} \right), \tag{2.16}$$

$$\left\langle A_{mk}^1 \mid H \mid B_{m'k}^2 \right\rangle = -\gamma_4 h \left(t_{3k}(m)\delta_{m-1,m'} + q\delta_{m,m'} \right), \tag{2.17}$$

$$\left\langle A_{mk}^1 \mid H \mid A_{m'k}^1 \right\rangle = \left\langle A_{mk}^2 \mid H \mid A_{mk'}^2 \right\rangle = E_A \delta_{m,m'}, \tag{2.18}$$

$$\left\langle B_{m\mathbf{k}}^1 \mid H \mid B_{m'\mathbf{k}}^1 \right\rangle = \left\langle B_{m\mathbf{k}}^2 \mid H \mid B_{m\mathbf{k}'}^2 \right\rangle = E_B \delta_{m,m'} \qquad (2.19)$$

where:

$t_{1\mathbf{k}}(m)$ and q are shown in Equation 2.10

$t_{2\mathbf{k}}(m)$ and $t_{3\mathbf{k}}(m)$ are expressed as

$$t_{2\mathbf{k}}(m) = \exp\left\{ i\left[-(k_x b/2) - \left(\sqrt{3}k_y b/2\right) + \pi\Phi(m-1+3/6) \right] \right\}$$

$$+\exp\left\{ i\left[-(k_x b/2) + \left(\sqrt{3}k_y b/2\right) - \pi\Phi(m-1+3/6) \right] \right\};$$

$$\qquad\qquad (2.20)$$

$$t_{3\mathbf{k}}(m) = \exp\left\{ i\left[-(k_x b/2) - \left(\sqrt{3}k_y b/2\right) + \pi\Phi(m-1+5/6) \right] \right\}$$

$$+\exp\left\{ i\left[-(k_x b/2) + \left(\sqrt{3}k_y b/2\right) - \pi\Phi(m-1+5/6) \right] \right\}.$$

It should be noted that the calculations based on the effective-mass approximation have trouble with the infinite order of the Hamiltonian matrix induced by the significant interlayer hopping integral of γ_3 [29,30]; this divergence also exists for ABC-stacked systems [264]. Nevertheless, through a qualitative perturbation analysis of γ_3 and other interlayer interactions, the minimal model, which regards γ_0 and γ_1 as the unperturbed terms, describes effectively the low-energy dispersions in the vicinity of the vertical edges in the first Brillouin zone. The generalized Peierls tight-binding model, which retains all important atomic interactions and magnetic fields, however, can provide comprehensive descriptions for graphites that offset the limitations of accuracy at low energies.

2.1.3 Rhombohedral Graphite

For ABC-stacked graphite, called rhombohedral graphite, the unit cell is chosen along the z-direction (Figure 2.1c). There are six atoms in a unit cell. The interlayer atomic interactions, based on the SWM model, take into account the nearest-neighbor intralayer interaction $\beta_0 = -2.73$ eV and five interlayer interactions $\beta_1 = 0.32$ eV, $\beta_2 = -0.0093$ eV, $\beta_3 = 0.29$ eV, $\beta_4 = 0.15$ eV and $\beta_5 = 0.0105$ eV, in which the former two refer to vertical atoms and the latter three are non-vertical [111]. The Hamiltonian matrix can be expressed as a combination of nine 2×2 submatrices for simplicity:

$$H_{ABC} = \begin{pmatrix} H_1 & H_2 & H_2^\star \\ H_2^\star & H_1 & H_2 \\ H_2 & H_2^\star & H_1 \end{pmatrix}, \qquad (2.21)$$

Where H_1 and H_2 take the forms

$$H_1 = \begin{pmatrix} 0 & \beta_0 f\left(k_x,k_y\right) \\ \beta_0 f^*\left(k_x,k_y\right) & 0 \end{pmatrix}; \tag{2.22}$$

$$H_2 = \begin{pmatrix} \left(\beta_4 \exp\left(ik_z I_z\right)+\beta_5 \exp\left(-i2k_z I_z\right)\right) f^*\left(k_x,k_y\right) \\ \left(\beta_3 \exp\left(ik_z I_z\right)+\beta_5 \exp\left(-i2k_z I_z\right)\right) f\left(k_x,k_y\right) \end{pmatrix}$$

$$\begin{matrix} \beta_1 \exp\left(ik_z I_z\right)+\beta_2 \exp\left(i2k_z I_z\right) \\ \left(\beta_4 \exp\left(ik_z I_z\right)+\beta_5 \exp\left(-i2k_z I_z\right)\right) f^*\left(k_x,k_y\right) \end{matrix}. \tag{2.23}$$

It is also noted that the hexagonal unit cell used here is not the primitive unit cell of rhombohedral graphite. The primitive unit cell should be a rhombohedral form that consists of two atoms and is inclined to the z-axis by an angle $\theta = \tan^{-1}\left(b/2/I_z\right)$ (Figure 2.1c). That is to say, the bases of the primitive unit cell are reduced from 6 to 2, as the rhombohedral unit cell is selected instead of the hexagonal one [110]. The Hamiltonian has an analytic solution near the zone edges H-K-H by using a continuum approximation [111,132,265]. This reflects the fact that the energy dispersions in the hexagonal cell can be zone-folded to the primitive rhombohedral one, and that the inversion symmetry is characterized similarly to that of simple hexagonal graphite. As a result, the physical properties of rhombohedral graphite might present certain features similar to those of monolayer graphene or simple hexagonal graphite, and their difference is only the degeneracy of energy states. A comparison between rhombohedral and hexagonal unit cells is made in detail in Chapter 5.

At $\mathbf{B} = B_0 \hat{z}$, the magnetically enlarged rectangle cell is chosen as the enlargement of the hexagonal unit cell along the z-axis for the convenience of calculations. Such a rectangular cell includes $3 \times 4R_B$ atoms and the Hamiltonian matrix elements are given by

$$\left\langle B_{mk}^1 \mid H \mid A_{m'k}^1 \right\rangle = \beta_0 \left(t_{1k}\left(m\right)\delta_{m,m'} + q\delta_{m,m'-1}\right), \tag{2.24}$$

$$\left\langle B_{mk}^2 \mid H \mid A_{m'k}^2 \right\rangle = \beta_0 \left(t_{3k}\left(m\right)\delta_{m,m'-1} + q\delta_{m,m'}\right), \tag{2.25}$$

$$\left\langle B_{mk}^3 \mid H \mid A_{m'k}^3 \right\rangle = \beta_0 \left(t_{3k}\left(m\right)\delta_{m,m'} + q\delta_{m,m'+1}\right), \tag{2.26}$$

$$\left\langle A_{mk}^1 \mid H \mid B_{m'k}^2 \right\rangle = \left\langle A_{mk}^2 \mid H \mid B_{m'k}^3 \right\rangle = \left\langle B_{mk}^1 \mid H \mid A_{m'k}^3 \right\rangle = \left(\beta_1 e^{i3k_z I_z} + \beta_2 e^{-6ik_z I_z}\right)\delta_{m,m'}, \tag{2.27}$$

$$\left\langle A^3_{m\mathbf{k}} \mid H \mid B^2_{m'\mathbf{k}} \right\rangle = \left(\beta_3 e^{ik_z I_z} + \beta_5 e^{-6ik_z I_z} \right) \left(t_{1\mathbf{k}}(m)\delta_{m,m'-1} + q\delta_{m,m'} \right), \qquad (2.28)$$

$$\left\langle A^2_{m\mathbf{k}} \mid H \mid B^1_{m'\mathbf{k}} \right\rangle = \left(\beta_3 e^{i3k_z I_z} + \beta_5 e^{-6ik_z I_z} \right) \left(t_{2\mathbf{k}}(m)\delta_{m,m'} + q\delta_{m,m'+1} \right), \qquad (2.29)$$

$$\left\langle A^1_{m\mathbf{k}} \mid H \mid B^3_{m'\mathbf{k}} \right\rangle = \left(\beta_3 e^{ik_z I_z} + \beta_5 e^{-6ik_z I_z} \right) \left(t_{3\mathbf{k}}(m)\delta_{m,m'-1} + q\delta_{m,m'} \right), \qquad (2.30)$$

$$\left\langle B^1_{m\mathbf{k}} \mid H \mid B^2_{m'\mathbf{k}} \right\rangle = \left(\beta_4 e^{i3k_z I_z} + \beta_5 e^{-6ik_z I_z} \right) \left(t_{2\mathbf{k}}(m)\delta_{m,m'} + q\delta_{m,m'+1} \right), \qquad (2.31)$$

$$\left\langle A^1_{m\mathbf{k}} \mid H \mid A^2_{m'\mathbf{k}} \right\rangle = \left\langle B^2_{m\mathbf{k}} \mid H \mid B^3_{m'\mathbf{k}} \right\rangle = \left(\beta_4 e^{-i3k_z I_z} + \beta_5 e^{6ik_z I_z} \right) \left(t_{3\mathbf{k}}(m)\delta_{m,m'-1} + q\delta_{m,m'} \right), \qquad (2.32)$$

$$\left\langle A^2_{m\mathbf{k}} \mid H \mid A^3_{m'\mathbf{k}} \right\rangle = \left(\beta_4 e^{-i3k_z I_z} + \beta_5 e^{6ik_z I_z} \right) \left(t_{2\mathbf{k}}(m)\delta_{m,m'} + q\delta_{m,m'+1} \right), \qquad (2.33)$$

$$\left\langle B^3_{m\mathbf{k}} \mid H \mid B^1_{m'\mathbf{k}} \right\rangle = \left(\beta_4 e^{-i3k_z I_z} + \beta_5 e^{6ik_z I_z} \right) \left(t_{2\mathbf{k}}(m)\delta_{m,m'} + q\delta_{m,m'+1} \right); \qquad (2.34)$$

$$\left\langle A^3_{m\mathbf{k}} \mid H \mid A^1_{m'\mathbf{k}} \right\rangle = \left(\beta_4 e^{-i3k_z I_z} + \beta_5 e^{6ik_z I_z} \right) \left(t_{1\mathbf{k}}(m)\delta_{m,m'} + q\delta_{m,m'+1} \right). \qquad (2.35)$$

The independent phase terms $t_{1\mathbf{k}}(m)$, $t_{2\mathbf{k}}(m)$, $t_{3\mathbf{k}}(m)$ and q are shown in Equations 2.10 and 2.20. The generalized tight-binding model, accompanied by an exact diagonalization method, can further be applied to study other physical properties, such as the optical absorption spectra [18,56–58,88,137,138] and plasma excitations [59,61,382]. Different kinds of external fields, for example, a modulated magnetic field [174], a periodic electric potential [175] and even a composite field [176], could also be included in the calculations simultaneously. Furthermore, this model can be applied to other layered materials with a precisely chosen layer sequence, such as graphene, MOS_2 and silicene, germanene, tinene and phosphorene [177–181]. The electronic structures and characteristics of wave functions are well depicted and the results are accurate and reliable within a wide energy range.

2.1.4 Gradient Approximation for Optical Properties

When graphite is subjected to an electromagnetic field, the optical spectral function $A(\omega)$ is used to describe its optical response. At zero temperature, $A(\omega)$ is expressed as follows according to the Kubo formula:

$$A(\omega) \propto \sum_{n^v, n^c} \int_{1stBZ} \frac{d\mathbf{k}}{(2\pi)^2} \left| \left\langle \Psi_{\mathbf{k}}^c (n^c) \left| \frac{\hat{\mathbf{E}} \cdot \mathbf{P}}{m_e} \right| \Psi_{\mathbf{k}}^v (n^v) \right\rangle \right|^2$$

$$\times Im \left\{ \frac{f\left[E_{\mathbf{k}}^c (n^c)\right] - f\left[E_{\mathbf{k}}^v (n^v)\right]}{E_{\mathbf{k}}^c (n^c) - E_{\mathbf{k}}^v (n^v) - \omega - I\gamma} \right\},$$

(2.36)

where the definite integral spans over the first Brillouin zone (1st BZ), $\hat{\mathbf{E}}$ is the direction of electric polarization, $\mathbf{P}$ is the momentum operator, $f[E_{\mathbf{k}}(n)]$ is the Fermi-Dirac distribution, m_e is the electron mass and γ is the phenomenological broadening parameter. $\hat{\mathbf{E}}$, which lies on the (x, y) plane, is chosen for a model study. $n^{c,v}$ is the energy band index measured from the Fermi level at zero field, or it represents the quantum number of each LS. The integration for all wave vectors is made within a hexahedron (a rectangular parallelepiped) in a zero (non-zero) magnetic field. The initial and final state satisfy the condition of $\Delta \mathbf{k} = 0$, which is responsible for the zero momentum of photons. This implies that only the vertical transitions are available in the valence and conduction bands. Using the gradient approximation [18,57], the velocity matrix element is evaluated from

$$M_{\mathbf{k}}^{c,v} (n^c, n^v) = \left\langle \Psi_{\mathbf{k}}^c (n^c) \left| \frac{\hat{\mathbf{E}} \cdot \mathbf{P}}{m_e} \right| \Psi_{\mathbf{k}}^v (n^v) \right\rangle \sim \left\langle \Psi_{\mathbf{k}}^c (n^c) \left| \partial_{k_y} H \right| \Psi_{\mathbf{k}}^v (n^v) \right\rangle \quad \text{for} \quad \hat{\mathbf{E}} \parallel y.$$

(2.37)

According to Equations 2.1, 2.5, 2.11 and 2.21, the velocity matrix elements of simple hexagonal, Bernal and rhombohedral graphite are given by

$$M_{AA}^{c,v} (n^c, n^v, \mathbf{k})$$

$$= \begin{pmatrix} C_{A,\mathbf{k}}^c & C_{B,\mathbf{k}}^c \end{pmatrix} \begin{pmatrix} 0 & (\alpha_0 + \alpha_3 h) \partial_{k_y} f \\ (\alpha_0 + \alpha_3 h) \partial_{k_y} f^* & 0 \end{pmatrix} \begin{pmatrix} C_{A,\mathbf{k}}^v \\ C_{B,\mathbf{k}}^v \end{pmatrix},$$

(2.38)

$$M_{AB}^{c,v}(n^c, n^v, \mathbf{k})$$

$$= \left(C_{A^1,\mathbf{k}}^c \quad C_{B^1,\mathbf{k}}^c \quad C_{A^2,\mathbf{k}}^c \quad C_{B^2,\mathbf{k}}^c \right)^*$$

$$\begin{pmatrix} 0 & \gamma_0 \partial_{k_y} f & 0 & \gamma_4 h \partial_{k_y} f^* \\ \gamma_0 \partial_{k_y} f^* & 0 & \gamma_4 h \partial_{k_y} f^* & \gamma_3 h \partial_{k_y} f \\ 0 & \gamma_4 h \partial_{k_y} f & 0 & \gamma_0 \partial_{k_y} f^* \\ \gamma_4 h \partial_{k_y} f & \gamma_3 h \partial_{k_y} f^* & \gamma_0 \partial_{k_y} f & 0 \end{pmatrix} \begin{pmatrix} C_{A^1,\mathbf{k}}^v \\ C_{B^1,\mathbf{k}}^v \\ C_{A^2,\mathbf{k}}^v \\ C_{B^2,\mathbf{k}}^v \end{pmatrix} \% ,$$

(2.39)

$$M_{ABC}^{c,v}\left(n^c, n^v, \mathbf{k}\right)$$

$$= \left(C_{A^1,\mathbf{k}}^c \quad C_{B^1,\mathbf{k}}^c \quad C_{A^2,\mathbf{k}}^c \quad C_{B^2,\mathbf{k}}^c \quad C_{A^3,\mathbf{k}}^c \quad C_{B^3,\mathbf{k}}^c \right)^*$$

$$\begin{pmatrix} 0 & \beta_0 \partial_{k_y} f & \beta_5' \partial_{k_y} f^* & 0 & \beta_5'^* \partial_{k_y} f & 0 \\ \beta_0 \partial_{k_y} f^* & 0 & \beta_5' \partial_{k_y} f & \beta_5' \partial_{k_y} f^* & \beta_5'^* \partial_{k_y} f^* & \beta_5'^* \partial_{k_y} f \\ \beta_5'^* \partial_{k_y} f & 0 & 0 & \beta_0 \partial_{k_y} f & \beta_5' \partial_{k_y} f^* & 0 \\ \beta_5'^* \partial_{k_y} f^* & \beta_5' \partial_{k_y} f & \beta_0 \partial_{k_y} f^* & 0 & \beta_5' \partial_{k_y} f & \beta_5' \partial_{k_y} f^* \\ \beta_5' \partial_{k_y} f^* & 0 & \beta_5'^* \partial_{k_y} f & 0 & 0 & \beta_0 \partial_{k_y} f \\ \beta_5' \partial_{k_y} f & \beta_5' \partial_{k_y} f^* & \beta_5'^* \partial_{k_y} f^* & \beta_5'^* \partial_{k_y} f & \beta_0 \partial_{k_y} f^* & 0 \end{pmatrix} \begin{pmatrix} C_{A^1,\mathbf{k}}^v \\ C_{B^1,\mathbf{k}}^v \\ C_{A^2,\mathbf{k}}^v \\ C_{B^2,\mathbf{k}}^v \\ C_{A^3,\mathbf{k}}^v \\ C_{B^3,\mathbf{k}}^v \end{pmatrix} ,$$

(2.40)

where $\partial_{k_y} f = \partial f\left(k_x, k_y\right) / \partial k_y = -\left(\sqrt{3}b/2\right)\exp\left(ibk_x/2\right)\sin\left(\sqrt{3}bk_y/2\right)$ and $\beta_5' = \beta_5 \exp(-i2k_z I_z)$. Substituting these matrix elements into $A(\omega)$ in Equation 2.36, and integrating all the available transitions over the first Brillouin zone and the quantum numbers, the spectral absorption function $A(\omega)$ is obtained. In addition, the absorption spectra are almost independent of the polarization direction when $\hat{\mathbf{E}}$ is on the x–y plane.

The velocity matrix significantly depends on the relation between the initial- and final-state wave functions, a main factor in determining the transition intensity and the optical selection rule. In the absence of external fields, what should be especially noted is the optical transitions centered around the highly symmetric $\mathbf{k}$ points, for example, Γ, M, K …, where the joint density of states (JDOS) and $M^{c,v}(n_c, n_v, \mathbf{k})$ have relatively large values. Under a magnetic field, the Bloch function at a fixed k_z is a linear combination of the products of the subenvelope function and the tight-binding function on each sublattice site in the enlarged unit cell (Equation 2.4). Consequently, $M^{c,v}(n_c, n_v, \mathbf{k})$ in Equation 2.37 can be expressed as

$$M^{c,v}(n^c, n^v, \mathbf{k}) \sim \left\langle \Psi_{\mathbf{k}}^c(n^c) \middle| \partial_{k_y} H \middle| \Psi_{\mathbf{k}}^v(n^v) \right\rangle$$

$$= \sum_{l,l'=1} \sum_{m,m'=1} \left(A_o^{l,c*} A_o^{l',v} + A_e^{l,c*} A_e^{l',v} \right) \left\langle A_{mk}^{l,c} \middle| \partial_{k_y} H \middle| A_{m'\mathbf{k}}^{l',v} \right\rangle$$

$$+ \left(A_o^{l,c*} B_o^{l',v} + A_e^{l,c*} B_e^{l',v} \right) \left\langle A_{mk}^{l,c} \middle| \partial_{k_y} H \middle| B_{m'\mathbf{k}}^{l',v} \right\rangle + \left(B_o^{l,c*} A_o^{l',v} + B_e^{l,c*} A_e^{l',v} \right) \left\langle B_{mk}^{l,c} \middle| \partial_{k_y} H \middle| A_{m'\mathbf{k}}^{l',v} \right\rangle$$

$$+ \left(B_o^{l,c*} B_o^{l',v} + A_e^{l,c*} A_e^{l',v} \right) \left\langle B_{mk}^{l,c} \middle| \partial_{k_y} H \middle| B_{m'\mathbf{k}}^{l',v} \right\rangle),$$

$$(2.41)$$

where the operator $\partial_{k_y} H$ takes the following forms:
For simple hexagonal graphite,

$$\left\langle A_{mk} \middle| \partial_{k_y} H \middle| B_{m'\mathbf{k}} \right\rangle = (\alpha_0 + \alpha_3 h) u_{1k}(m) \delta_{m,m'}, \qquad (2.42)$$

For Bernal graphite,

$$\left\langle B_{mk}^1 \middle| \partial_{k_y} H \middle| A_{m'\mathbf{k}}^1 \right\rangle = -\gamma_0 u_{1k}(m) \delta_{m,m'}, \qquad (2.43)$$

$$\left\langle B_{mk}^1 \middle| \partial_{k_y} H \middle| A_{m'\mathbf{k}}^2 \right\rangle = \gamma_4 h u_{1k}(m) \delta_{m,m'}, \qquad (2.44)$$

$$\left\langle B_{mk}^2 \middle| \partial_{k_y} H \middle| B_{m'\mathbf{k}}^1 \right\rangle = \gamma_3 h u_{2k}(m) \delta_{m,m'}, \qquad (2.45)$$

$$\left\langle A_{mk}^2 \middle| \partial_{k_y} H \middle| B_{m'\mathbf{k}}^2 \right\rangle = -\gamma_0 u_{3k}(m) \delta_{m-1,m'}, \qquad (2.46)$$

$$\left\langle A_{mk}^1 \middle| \partial_{k_y} H \middle| B_{m'\mathbf{k}}^2 \right\rangle = -\gamma_4 h u_{3k}(m) \delta_{m-1,m'}, \qquad (2.47)$$

For rhombohedral graphite,

$$\left\langle B_{mk}^1 \middle| \partial_{k_y} H \middle| A_{m'\mathbf{k}}^1 \right\rangle = \beta_0 u_{1k}(m) \delta_{m,m'}, \qquad (2.48)$$

$$\left\langle B_{mk}^2 \middle| \partial_{k_y} H \middle| A_{m'\mathbf{k}}^2 \right\rangle = \beta_0 u_{3k}(m) \delta_{m,m'-1}, \qquad (2.49)$$

$$\left\langle B_{mk}^3 \middle| \partial_{k_y} H \middle| A_{m'\mathbf{k}}^3 \right\rangle = \beta_0 u_{3k}(m) \delta_{m,m'}, \qquad (2.50)$$

$$\left\langle A_{mk}^3 \middle| \partial_{k_y} H \middle| B_{m'\mathbf{k}}^2 \right\rangle = \left(\beta_3 e^{ik_z l_z} + \beta_5 e^{-6ik_z l_z} \right) u_{1k}(m) \delta_{m,m'-1}, \qquad (2.51)$$

$$\left\langle A_{m\mathbf{k}}^{2}\left|\partial_{k_y}H\right|B_{m'\mathbf{k}}^{1}\right\rangle = \left(\beta_3 e^{i3k_zl_z} + \beta_5 e^{-6ik_zl_z}\right)u_{2\mathbf{k}}(m)\delta_{m,m'},\qquad(2.52)$$

$$\left\langle A_{m\mathbf{k}}^{1}\left|\partial_{k_y}H\right|B_{m'\mathbf{k}}^{3}\right\rangle = \left(\beta_3 e^{ik_zl_z} + \beta_5 e^{-6ik_zl_z}\right)u_{3\mathbf{k}}(m)\delta_{m,m'-1},\qquad(2.53)$$

$$\left\langle B_{m\mathbf{k}}^{1}\left|\partial_{k_y}H\right|B_{m'\mathbf{k}}^{2}\right\rangle = \left(\beta_4 e^{i3k_zl_z} + \beta_5 e^{-6ik_zl_z}\right)u_{2\mathbf{k}}(m)\delta_{m,m'},\qquad(2.54)$$

$$\left\langle A_{m\mathbf{k}}^{1}\left|\partial_{k_y}H\right|A_{m'\mathbf{k}}^{2}\right\rangle = \left\langle B_{m\mathbf{k}}^{2}\left|\partial_{k_y}H\right|B_{m'\mathbf{k}}^{3}\right\rangle = \left(\beta_4 e^{-i3k_zl_z} + \beta_5 e^{6ik_zl_z}\right)u_{3\mathbf{k}}(m)\delta_{m,m'-1},\quad(2.55)$$

$$\left\langle A_{m\mathbf{k}}^{2}\left|\partial_{k_y}H\right|A_{m'\mathbf{k}}^{3}\right\rangle = \left(\beta_4 e^{-i3k_zl_z} + \beta_5 e^{6ik_zl_z}\right)u_{2\mathbf{k}}(m)\delta_{m,m'},\qquad(2.56)$$

$$\left\langle B_{m\mathbf{k}}^{3}\left|\partial_{k_y}H\right|B_{m'\mathbf{k}}^{1}\right\rangle = \left(\beta_4 e^{-i3k_zl_z} + \beta_5 e^{6ik_zl_z}\right)u_{2\mathbf{k}}(m)\delta_{m,m'},\qquad(2.57)$$

$$\left\langle A_{m\mathbf{k}}^{3}\left|\partial_{k_y}H\right|A_{m'\mathbf{k}}^{1}\right\rangle = \left(\beta_4 e^{-i3k_zl_z} + \beta_5 e^{6ik_zl_z}\right)u_{1\mathbf{k}}(m)\delta_{m,m'},\qquad(2.58)$$

where $u_{i\mathbf{k}}(m)$ is the partial derivative of $t_{i\mathbf{k}}(m)$ with respect to k_y and is expressed as follows:

$$u_{1\mathbf{k}}(m) = -\left(\sqrt{3}b/2\right)\exp\left\{i\left[-(k_xb/2)-\left(\sqrt{3}k_yb/2\right)+\pi\Phi(m-1+1/6)\right]\right\}$$
$$+\left(\sqrt{3}b/2\right)\exp\left\{i\left[-(k_xb/2)+\left(\sqrt{3}k_yb/2\right)-\pi\Phi(m-1+1/6)\right]\right\},$$

$$u_{2\mathbf{k}}(m) = -\left(\sqrt{3}b/2\right)\exp\left\{i\left[-(k_xb/2)-\left(\sqrt{3}k_yb/2\right)+\pi\Phi(m-1+3/6)\right]\right\}$$
$$+\left(\sqrt{3}b/2\right)\exp\left\{i\left[-(k_xb/2)+\left(\sqrt{3}k_yb/2\right)-\pi\Phi(m-1+3/6)\right]\right\};\qquad(2.59)$$

$$u_{3\mathbf{k}}(m) = -\left(\sqrt{3}b/2\right)\exp\left\{i\left[-(k_xb/2)-\left(\sqrt{3}k_yb/2\right)+\pi\Phi(m-1+5/6)\right]\right\}$$
$$+\left(\sqrt{3}b/2\right)\exp\left\{i\left[-(k_xb/2)+\left(\sqrt{3}k_yb/2\right)-\pi\Phi(m-1+5/6)\right]\right\}.$$

Nevertheless, $M^{c,\,v}(n^c, n^v, \mathbf{k})$ is simplified as the product of three matrices: the operator $\partial_{k_y}H$ and the eigenfunctions of the initial and final states. It can be deduced as a simple inner product of the subenvelope functions, due to the fact that the Peierls phase slowly changes in the enlarged unit cell so that this derivative term $\partial H/\partial k_y$ can be taken out of the summation in Equation 2.36. Considering both interlayer and intralayer atomic interactions, one finds that while all the hopping integrals, α's, β's or γ's, make contributions to the absorption spectrum, it is the relatively stronger in-plane atomic interactions, α_0, β_0 or γ_0, that play the most important role in the optical transitions. When the occupied LSs are excited to the unoccupied ones, the available excitation channels satisfy the general selection rule, $\Delta n = n^c_{A^l(B^l)} - n^v_{B^l(A^l)} = \pm 1$, where $n^l_{A(B)}$ indicates the quantum mode for the A(B) sublattices on the *l*-th layer. The detailed calculation results are discussed in the following chapters.

2.2 Scanning Tunneling Microscopy and Spectroscopy

STM has been developed into the most important instrument for imaging the surface structure since its invention by Binnig and Rohrer in 1982 [182]. By combining a metallic tip and a precise scanning device, both the lateral and vertical surface information can be obtained with atomic resolutions by means of measuring the tunneling currents through the tip and samples. Obtained from the selected positions of examined materials are the full surface topographies, such as the bond lengths [183], crystal orientations [128,130], corrugation [184], edge structures [185–188], local vacancies [189,190], dislocations [191,192] and defects [193,194]. The spectroscopy consists of a scanning tip, high-voltage (HV) amplifiers, piezoelectric devices and a computer, as shown in Figure 2.2. Based on the quantum tunneling effect, a small electric current is built through the sample and the tip by applying a bias voltage [203]. The tunneling current is exponentially decayed with the increasing surface-tip distance, flowing from the occupied electronic states of the tip into the unoccupied ones of the surface under a positive bias voltage V > 0, or vice versa. The spectroscopic response is barely influenced by the background in an ultrahigh vacuum environment.

In general, the quantum tunneling current is served as a feedback signal in the piezoelectric feedback device to control the tip movement in z-direction; meanwhile, the voltage ramp is used to control the x–y scanner. STM is operated in the most commonly used mode, namely the constant current mode, in which the feedback electronics adjust the tip-surface distance to maintain a constant tunneling current/voltage by tuning the applied voltage of the z-piezo. Therefore, a map of the surface topography can be determined when the tip scans over the sample surface. Up to now, the atomic resolution of STM measurements have reached the order of ~1Å for lateral and ~0.1Å for vertical surfaces [128,130,204–206], mainly due to the significant dependence of the tunneling current on the surface-tip distance. On the other hand, for magnetic materials, the spin-polarized current can be resolved by means of a ferromagnetic/antiferromagnetic probe tip in spin-polarized STM. This spin-dependent spectroscopic mode was first proposed by Pierce in 1988 and realized by Wiesendanger et al. in 1990 [195,196], who provide detailed

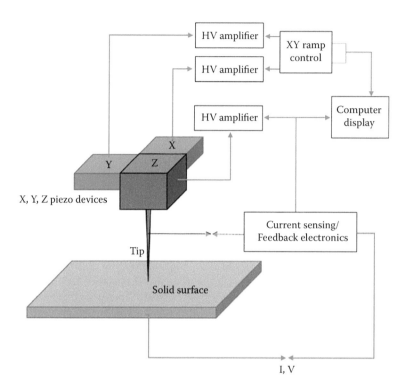

FIGURE 2.2

Schematic diagram of STM. The feedback electronics and the HV-amplifier control z-movement via the tunneling current I, while the voltage ramp is used for xy-movement.

information of the spin polarizations in the magnetic domain of the sample surface.

STM has been widely used to reveal the spatially atomic distributions of carbon-related systems from 1D to 3D, for example, graphites [64,65,183,184], graphenes [128–130,190,194,206], graphene ribbons [185–188] and carbon nanotubes. The nanoscale precision in measurements is helpful in understanding the unique geometric structures, and in illustrating the relative positions of hexagonal lattices under a specified stacking configuration. The nonchiral, armchair, zigzag, and chiral structures can also be identified for the cross-sections of carbon nanotubes and the finite-sized edges of ribbons. In addition, the geometric topographies of low-dimensional systems can also be characterized by other experimental techniques, such as transmission electron microscopy (TEM) [48], scanning transmission electron microscope (STEM) [197,198], atomic force microscopy (AFM) [199,200] and low-energy electron diffraction (LEED) [201,202].

STS, an extension of STM, is used to demonstrate the tunneling current through the tip-surface junction in the constant height mode [203]. Sweeping over the bias voltage V in the absence of feedback, STS is able to characterize the electronic properties of the selected conducting surfaces, based on the measured I-V curves and differential conductance dI/dV. When the density of states (DOS) of the tip is assumed to be constant, the tunneling current I is decided by the total electron transitions from the sample to the tip, with energies from Fermi level to eV. In this way, the normalized differential conductance, defined as (dI/dV)/(I/V), is interpreted as proportional to the DOS at energy eV with extra V-dependent integral constants. In general, the normalized differential conductance is measured by using a lock-in amplifier in a small range of AC modulation of dV; this technique can greatly reduce the noise of the measured conductance. The current experiment resolution reaches up to 10 pA. The STS measurements directly examine the diverse electronic properties of graphite-related systems, such as the verification of the geometry-dependent Van Hove singularities (VHSs) in graphites [64,65], graphenes [83,84,87,128,129], graphene nanoribbons [164–167] and carbon nanotubes [142,207,208]. On the other hand, STS is also able to identify the magneto-electronic properties of graphite-related systems [81–84]. Specifically, in terms of simultaneous STM and STS measurements on the same sample, the relationship between the geometric and electronic properties gives the width-dependent energy gaps and standing wave functions of graphene nanoribbons [164–167], the chiral-dependent gaps of carbon nanotubes [142,207,208] and the stacking-dependent band structures of graphites [65] and graphenes [87,129].

2.3 Angle-Resolved Photoemission Spectroscopy

ARPES is a powerful tool to directly study the energy dispersions of the occupied electronic states in the Brillouin zone [50,51,71–75,78,79,232]. When the sample is illuminated by soft x-rays (Figure 2.3), the electrons stimulated by the incident photons may escape from the material due to the photoelectric effect, and hence provide the image of the underlying electronic structures of materials.

The kinetic energy of photoelectrons is given by the conservation of energy principle. By an angle-resolved energy analyzer, the photoelectrons are collected in a solid angle Ω which derives from a polar angle θ and an azimuthal angle φ. Evaluated from the free particle model, the parallel component of the momentum is conserved due to the translational symmetry, while the perpendicular component momentum is not conserved due to the breaking of translation symmetry along the normal direction. In general, the sample is synthesized in a portable chamber for conveniently measuring the in-situ electronic structure. ARPES is mainly focused on 2D or quasi-2D materials where the energy dispersions perpendicular to the surface are negligible.

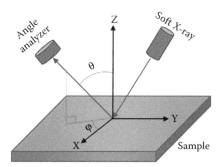

FIGURE 2.3

Schematic diagram of ARPES. Soft x-ray is used as the light source, and θ and φ are used to describe the polar and azimuthal angles of the detected photoelectrons, respectively.

However, the non-conservation issue in the normal direction might be overcome by means of mapping the important characteristics to the band structure at certain k_z points, for example, the successful identification of the 3D band structures of graphite [71–75]. Specifically, the ARPES measurements can provide an insight into the electron–electron [71] and electron–phonon [209] interactions in many-body systems. Improvements of experimental resolutions in energy and momentum have become a critical factor for investigating low-dimensional materials.

2.4 Absorption Spectroscopy

Optical spectroscopy is useful to characterize the absorption [10,87,89,150,244,255], transmission [24,93,95,246,248] and reflectivity [9,22,94–96] of a variety of low-dimensional solid-state materials. Absorption spectroscopy, based on the analytical technique of measuring the fraction of incident radiation absorbed by a sample, is one of the most versatile and widely used techniques in physics and chemistry. Operated over a range of frequencies, absorption spectroscopy is employed as an analytical tool to determine the intensity of the absorbed radiation which varies as a function of frequency for characterizing the optical and electronic properties of materials. The experimental setup relating to the light source, sample arrangement and detection technique varies significantly depending on the frequency range and the purpose of the experiment.

The most common setup of the apparatus is to emit a radiation beam onto the sample and to detect the transmitted intensity of the radiation that can be used to calculate the absorption, as shown in Figure 2.4. Optical experiments are usually performed with a broadband light source, for which the intensity and frequency can be adjusted in a broad range. The most-utilized

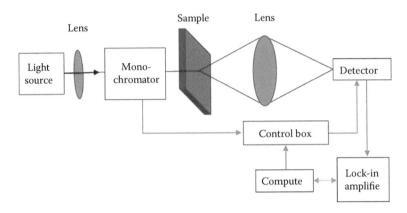

FIGURE 2.4
Schematic diagram of optical absorption spectroscopy. The spectrum is performed in a broad-band range from visible to ultraviolet. The monochromator is used to select the desired frequency range of the light source.

light sources can be classified into three kinds according to the operated frequency ranges: (1) the xenon-mercury arc lamp, under high-pressure in the far-infrared region [210,211]; (2) the black-body source of a heated SiC element in the mid- to near-infrared spectral range [212–214]; and (3) the tungsten halogen lamp in a continuous spectrum from visible to near-infrared to near-ultraviolet [24,95,215,216], which is commonly adopted for the analytical characterization of the optical properties of materials over a broad spectrum because it can be operated at a high temperature (>3000 K) under the inert-halogen mixture atmosphere.

Absorption spectra can be measured using a Fourier-transform spectrometer [217,218], linear photodiode array spectrophotometer [219–223,255,256] or charge-coupled device spectrometer [224–226]. The Fourier-transform spectrometer is based on measuring the coherence of electromagnetic radiations, in either time-domain or space-domain electromagnetic radiation, for example, the Bruker IFS125 spectrometer, which has resolved linewidths <0.001 cm [217,218]. On the other hand, the photodiode array spectrophotometer, consisting of hundreds of linear high-speed detectors integrated on a single chip, simultaneously measures the dispersive light over a wide frequency range. According to Fermi's golden rule, the electronic properties can be deduced from the optical spectra, in which the absorption peaks have widths and shapes that are primarily determined by the transition probability and the DOS. For graphite-related materials, the frequencies not only depend on the dimensionality, atomic interactions and boundary conditions of the systems, but also on the external electric and magnetic fields.

Magneto-absorption spectroscopy can also be used to study the quantization phenomena of low-dimensional systems, such as the quasi-LLs in graphene ribbons [172,173], AB-effect in carbon nanotubes [149], LLs in few-layer graphenes and [20–24] LSs in bulk graphites [89–94,96]. The magnetic field can be generated by a superconducting magnet; a 25-T cryogen-free superconducting magnet has been developed at the High Field Laboratory [227,228]. Furthermore, ultrahigh magnetic fields can be generated by using a semi-destructive single-turn coil technique that provides a pulsed field >100 T for pulse lengths of ten µs [229,255,256]. In addition, magneto-Raman spectroscopy also provides a convenient and powerful approach for tailoring the magneto-electronic properties based on the Raman scattering principle [139,262,263].

3

Simple Hexagonal Graphite

AA-stacked graphite possesses the highest stacking symmetry among the layered graphites. The hexagonal symmetry, AA-stacking configuration and significant interlayer atomic interactions are responsible for the unusual essential properties. The non-titled Dirac-cone structure is formed along the k_z-direction, in which its width is more than 1 eV. The 3D Dirac cone covers free electrons and holes with the same density, leading to semimetallic behavior with an obvious plateau structure in low-energy density of states (DOS). It is further quantized into 1D parabolic Landau subbands (LSs) without any crossings or anti-crossings. Each well-behaved LS contributes two asymmetric square-root-form peaks in DOS. A lot of LSs, which can cross the Fermi level, belong to the valence or conduction LSs. Specifically, this creates intraband and interband inter-LS magneto-optical excitation channels. The quantized energies have a simple dependence on $(B_0, n^{c, v}, k_z)$, so that the magneto-absorption spectra present beating features. Such phenomena are never predicted or observed in other condensed-matter systems. On the other hand, the zero-field absorption spectrum is largely suppressed and almost featureless at a low frequency because of many forbidden vertical transitions. AA-stacked graphite and graphenes are quite different from each other in terms of electronic and optical properties. The experimental verifications of energy bands, DOSs and absorption spectra of simple hexagonal graphite can be utilized to determine the critical intralayer and interlayer atomic interactions.

3.1 Electronic Structures without External Fields

The 2D π-electronic structure of a monolayer graphene is reviewed first. Given the interlayer atomic interactions $\alpha_1 = 0$, $\alpha_2 = 0$ and $\alpha_3 = 0$ in Equation 2.5, one can obtain the band structure of monolayer graphene, that is, $E^{c,v}(\mathbf{k}) = \pm\alpha_0 \mid f(k_x, k_y) \mid = \pm\alpha_0[1 + 4\cos(3bk_x/2)\cos(\sqrt{3}bk_y/2) + 4\cos^2(\sqrt{3}bk_y/2)]^{1/2}$. The band structure is simplified as the projection of energy dispersion of the simple hexagonal graphite on the $k_z = 0$ plane (the red hexagon in Figure 3.1a).

Both conduction and valence bands are symmetric around the Fermi level (E_F) along K→Γ→M→K. In the low-energy region, the energy dispersion is described by $E^{c,v} = \pm 3\alpha_0 bk/2$, which characterizes an isotropic Dirac cone

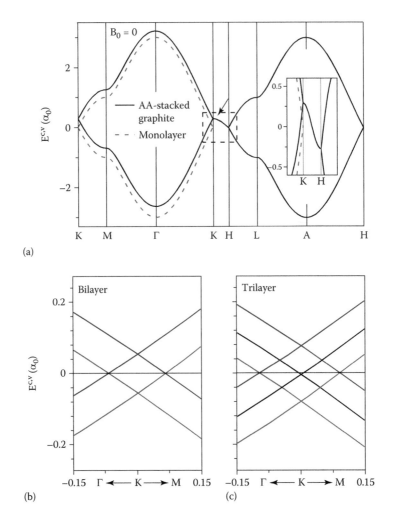

FIGURE 3.1
Band structures for AA-stacked (a) graphite and monolayer graphene, (b) bilayer systems and (c) trilayer systems.

centered at the K point (the Fermi level). There are special band structures at highly symmetric points in the first Brillouin zone (1st BZ), for example, the local maximum $E^c = 3\alpha_0$ and the local minimum $E^v = -3\alpha_0$ at the Γ point, and the saddle points $E^{c,v} = \pm\alpha_0$ at the M point. Such critical points in the energy-wave-vector space would induce Van Hove singularities in DOS. The bandwidth is evaluated as $6\alpha_0$, which is determined by the difference between the two local extreme values at the Γ point. Monolayer graphene is a zero-gap semiconductor with a vanishing DOS at E_F (Figure 3.2b), that is, free carriers are absent at zero temperature.

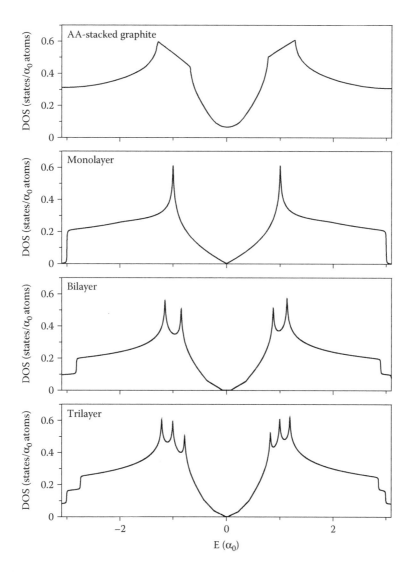

FIGURE 3.2
Density of states for (a) simple hexagonal graphite, and (b) monolayer, (c) bilayer and (d) trilayer graphenes.

Interlayer atomic interactions can dramatically change electronic structures. According to Equation 2.5, the energy dispersions of simple hexagonal graphite without a magnetic field are shown by the black curves in Figure 3.1a. There exists one pair of valence and conduction bands, in which the former is no longer symmetric to the latter around the Fermi level, E_F. Energy bands are highly anisotropic and strongly dependent on k_z. At a fixed k_z, the (k_x, k_y)-dependent energy dispersions resemble those of a monolayer graphene. Moreover, the critical points are very sensitive to the change of k_z, for example, those at the corners (K and H; Figure 2.1e), middle points between two corners (M and L) and centers of the k_x–k_y plane (Γ and A). The energy spacing of a monolayer-like band structure grows when k_z moves from K to H. That is, the Dirac-cone structures can survive and remain similar with the increase/decrease of k_z. This will be directly reflected in the magnetic quantization. The middle points, which correspond to the saddle points with high DOS, are expected to present the strong absorption spectra. Overall, the π-electronic width is evaluated as the energy difference between the maximum energy at the Γ point and the minimum energy at the A point: that is, $[2(\alpha_1+\alpha_2)+3(\alpha_0+2\alpha_3)] - [-2(\alpha_1-\alpha_2)-3(\alpha_0-2\alpha_3)]=4\alpha_1+6\alpha_0$.

In the low-energy approximation around the corners along the K–H direction, Equation 2.5, used to describe Dirac-type energy dispersions, can be expressed as

$$E_{\pm}^{c,v}\left(k_x,k_y,k_z\right)= E_D \pm v_F\left|k\right| \qquad (3.1)$$

where $E_D=\alpha_1 h+2\alpha_2(h^2/2-1)$, $v_F=3b(\alpha_0+\alpha_3 h)/2$ (the Fermi velocity) and $|k|=\sqrt{k_x^2+k_y^2}$. The first term E_D indicates the Dirac-point energy. The second term represents the conical energy dispersion, the slope of which particularly shows a slight discrepancy on k_z.

A closer examination is necessary to explore the dependence of the Dirac cone on k_z. Given by E_D in Equation 3.1, the localization of the Dirac point is described as a correspondence to the energy dispersion along the K–H line (indicated by the arrow in Figure 3.1). In the vicinity of the zone corners, the conduction and valence Dirac cones overlap with each other. At the K point, the state energy of the Dirac point, $2(\alpha_1+\alpha_2)$, is higher than E_F. This indicates that the valence states between E_F and the Dirac point are regarded as free holes in the low-lying valence bands. As the states gradually move away from K toward H, the carrier density of free holes decreases because the Dirac point gets lower. It is not until the Dirac point approaches the Fermi level that there are no free carriers. With a further increase of k_z ($E_D<E_F$), the free carriers change into electrons, a process that is determined by the Fermi level in the conduction Dirac cone. The electron density reaches a maximum value at the H point. In short, this means that the interlayer atomic interactions induce free-hole (free-electron) pockets in the low-energy valence (conduction) bands near the K (H) point. Two kinds of free carriers have the same density. Furthermore, the Dirac points of cone structures are located at the corners of the 1st BZ during the variation of k_z (Figure 2.1e). These are expected to play an important role in determining the essential physical properties, for example, the optical properties, magneto-electronic and magneto-optical properties, electronic excitations and transport properties.

In the case of 2D multilayer AA-stacked graphene, there are N pairs of valence and conduction Dirac cones, mainly because it has the highest stacking symmetry. For example, there are two and three pairs in bilayer and trilayer graphenes, respectively (Figure 3.1b and c), and the overlap of valence and conduction cones indicates semimetallic behavior. The Dirac-cone structures, which are initiated from the K point, are almost symmetric around the Fermi level. The Dirac-point energies between $-2\alpha_1$ and $2\alpha_1$ are described by [230].

$$E_D = 2\cos\left[j\pi / \left(N+1 \right) \right]\alpha_1 \qquad (3.2)$$

in the low-energy approximation (ignoring α_2 and α_3), where $j = 1, 2, \ldots N$. When N is an odd number, the Dirac point of the middle cone structure touches the Fermi level. With an increase of layer number, the multi-cone structures gradually evolve into a 3D structure with a significant k_z-dependent bandwidth. However, it might have certain important differences compared to AA-stacked few-layer graphenes and graphite in terms of the essential properties as a result of the confinement effect along the z-direction, for example, the optical threshold frequency, DOS and the features of magneto-absorption peaks.

On the experimental side, angle-resolved photoemission spectroscopy (ARPES) can directly identify the wave-vector-dependent energy bands. Using high-resolution ARPES measurements, the unusual, dimension-created electronic structures have been verified for carbon-related systems with hexagonal symmetry, including graphene nanoribbons, number- and stacking-dependent graphenes and AB-stacked graphite. The confirmed characteristics cover the confinement-induced energy gap and 1D parabolic bands in finite-width nanoribbons [163,232]; the Dirac-cone structure in monolayer graphene [78,233,234]; two/three pairs of linear bands in bilayer/trilayer AA stacking [50,51]; two pairs of parabolic bands in bilayer AB stacking; partially flat [78,79], sombrero-shaped and linear bands in trilayer ABC stacking [80]; and the bilayer- and monolayer-like energy dispersions in Bernal graphite at the K and H points, respectively [71–75]. The 3D band structure of AA-stacked graphite is worthy of detailed ARPES examinations, especially for Dirac-cone structures and the saddle points along the K–H and M–L lines, respectively. Such measurements can determine the intralayer and interlayer hopping integrals and their significant effects.

The primary characteristics of electronic structures directly reflect on the DOS. The low-energy special structures in DOS are dominated by the stacking configuration or the interlayer atomic interactions. In the range of $|E| \leq 0.3$ eV and $B_0 = 0$, simple hexagonal graphite presents a plateau structure centered around $E = 0$ (the Fermi level), as shown in Figure 3.2a. This originates from the superposition of all k_z-dependent Dirac-cone structures with various circular cross-sections in the (k_x, k_y) plane. A finite DOS at $E = 0$ clearly illustrates the semimetallic behavior. The DOS grows quickly with the increase of E. There exist two very cusp structures at the middle energies of $[-2(\alpha_1 - \alpha_2) + (\alpha_0 - 2\alpha_3)] \leq E \leq [2(\alpha_1 + \alpha_2) + (\alpha_0 + 2\alpha_3)]$ and $[-2(\alpha_1 - \alpha_2) - (\alpha_0 - 2\alpha_3)] \leq E \leq [2(\alpha_1 + \alpha_2) - (\alpha_0 + 2\alpha_3)]$ (Equation 2.6), mainly owing to the saddle points along the M–L line (Figure 2.1e) [231]. On the other hand, monolayer graphene exhibits a V-shape DOS near $E=0$, as shown in Figure 3.2b. The DOS vanishes at the Fermi level, leading to the semiconducting behavior. For N-even systems, the low-energy DOS corresponds to a plateau structure, for example, that of bilayer graphene (Figure 3.2c). However, it is a superposition of the plateau and V-shape structures for N-odd systems, such as the DOS of trilayer graphene (Figure 3.2d). Apparently, the AA-stacked graphenes of $N \geq 2$ belong to semimetals. At middle energy, the symmetric peaks of the logarithmic form mainly come from the saddle point (the M point in Figure 3.1a through c), in which their number is proportional to that of the layer (Figure 3.2b through d).

3.2 Optical Properties without External Fields

The main features of absorption spectra are determined by the velocity matrix element, carrier distribution and DOS. At low frequency, the velocity matrix element is just the Fermi velocity in AA-stacking systems, mainly owing to the similar Dirac cone with isotropic linear dispersions [231,241]. That is, AA-stacked graphite and multilayer graphenes have an identical excitation strength for each available channel. The former, as indicated in Figure 3.3a, presents a largely reduced low-frequency absorption spectrum and a shoulder structure at $\omega \sim 4\alpha_1 \sim 0.56\alpha_0$. For any given k_z, the vertical transitions are forbidden when half of the excitation frequency is smaller than the energy difference (E_{th}) between the Fermi-momentum state ($\mathbf{k}_F$) and the Dirac point. E_{th} is about $2\alpha_1$ for various k_z's, so the absorption spectrum is very weak at $\omega < 2\alpha_1$. With the increase of frequency, it exhibits a shoulder structure and grows quickly as the deeper or higher electronic states make more contributions. On the other hand, the absorption spectrum of monolayer graphene is linearly proportional to the excitation frequency (Figure 3.3b), directly reflecting the linear energy dependence of the DOS (Figure 3.2b). As for the middle-frequency absorption spectrum, graphite and graphene, respectively, exhibit a very prominent plateau and symmetric peak at $2\alpha_0 - 4\alpha_0 \leq \omega \leq 2\alpha_0 + 4\alpha_0$ and $\omega = 2\alpha_0$. The former originates from the saddle points along the M–L line (Equation 2.6), with a rather high DOS. Such structures are the so-called π-electronic absorption peaks, frequently observed in carbon-related systems with sp² bondings (discussed later).

A A-stacked layered graphenes present unusual low-frequency absorption spectra with the variation of layer number, as clearly indicated in Figure 3.3b through d. The critical factor is the well-behaved N pairs of Dirac cones that are almost symmetric around the Fermi level [241]. The wave functions of these cone structures are the symmetric or anti-symmetric linear superposition of the layer-dependent tight-binding functions (e.g., Equation 2.7),

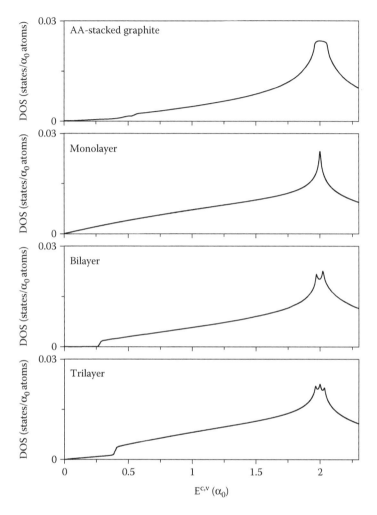

FIGURE 3.3
Optical absorption spectra of AA-stacked (a) graphite, and (b) monolayer, (c) bilayer and (d) trilayer graphenes.

leading the available excitation channels to only arise from the same Dirac cone. That is, the inter-Dirac-cone vertical transitions are absent. The N-odd systems have a zero-threshold frequency, as the Dirac point of the middle cone structure touches with the Fermi level, for example, the trilayer system (Figure 3.3d). However, the optical gaps, which are characterized by the energy spacing of the highest occupied and the lowest unoccupied Dirac points (Equation 3.2), are finite in the N-even systems. They decline with the increasing layer number, and the highest threshold frequency is $2\alpha_1$ for bilayer AA stacking (Figure 3.3c) [59]. As to the other intra-Dirac-cone excitations, their threshold spectra exhibit shoulder structures with an absorption frequency determined by the Fermi-momentum state (or about double that of the energy difference between the Dirac point and the Fermi level). In addition, absorption spectra might reveal two sub-shoulders because of the slightly asymmetric Dirac-cone structures due to the interlayer atomic inter-actions, for example, those of $N=4$ and 5 [241]. Specifically, the change of layer number results in the crossing behavior. AA-stacked graphenes and graphite possess almost identical low-frequency optical properties when N grows to 30 (detailed discussions in [241]). The dimension-induced impor-tant differences can be observed under the obvious confinement effect.

The aforementioned features of vertical excitation spectra can be veri-fied by optical spectroscopies, such as the absorption [9,10], transmission [9,20,21,23,24], reflection [9,22,94–96], Raman scattering [139,262,263] and Rayleigh scattering spectroscopies [278]. Experimental measurements have confirmed the rich and diverse optical properties of carbon-related systems, such as Bernal graphite [89–92], graphite intercalation compounds [248,249], layered graphenes [20–24], graphene nanoribbons [160,242], carbon nano-tubes [149,150] and carbon fullerenes [243,244]. Such systems possess the ~ 5–6 eV π peak arising from the 2pz-orbital bondings; that is, all the sp2-bonding systems can create this prominent peak. The AB- and ABC-stacked graphenes differ from each other in terms of absorption frequencies, spectral structures and electric-field-induced excitation spectra [86,87,131,255,272]. Moreover, carbon nanotubes exhibit a strong dependence of asymmetric absorption peaks on radius and chirality [149]. The important features in AA-stacked graphenes and graphite are worthy of systematic experimental investigations, especially for the dependence of optical gap, shoulder struc-ture, π peak and spectral intensity on the layer number.

3.3 Magnetic Quantization

3.3.1 Landau Levels and Wave Functions

In the presence of $\mathbf{B} = B_0\hat{z}$, electrons are flocked on the x–y plane to form transverse cyclotron motions, while the motion along the field direction remains intact. The 3D electronic states in simple hexagonal graphite are evolved into one group of so-called LSs, which are dispersed along the $\hat{k}_z$ direction, but highly degenerate on the k_x–k_y plane. This implies that the k_z-dependent LSs are directly quantized from the corresponding k_z-dependent Dirac cones along K–H in the absence of a magnetic field. The study of the magnetic quantization of a Dirac cone in monolayer graphene is the first step to realizing the magneto-electronic properties of graphites.

The electronic states of a Dirac cone are magnetically quantized into one group of valence and conduction Landau levels (LLs). Each LL, dispersionless

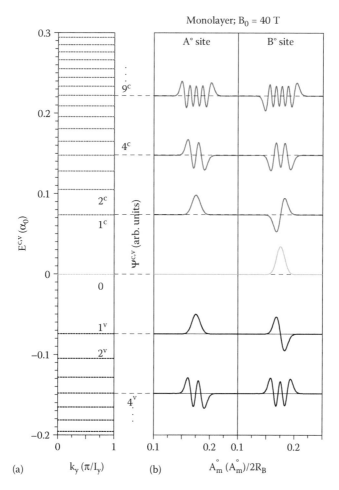

FIGURE 3.4
Landau levels of monolayer graphene at $B_0=40$ T: (a) energy spectrum and (b) amplitudes of subenvelope functions at two sublattices.

along k_x and k_y, is fourfold degenerate without the consideration of spin degeneracy. The occupied valence and unoccupied conduction LLs are symmetric around E_F, as shown in Figure 3.4a.

The quantum numbers, characterized by the zero-point values ones of the subenvelope functions, are indicated by n^c and n^v for the conduction and valence LLs, respectively. Considering the sequence of LLs, one finds that the $n^{c,v}=0$ LLs are located at E_F, and the $n^{c,v}=1, 2, 3, ...$ LLs are counted away from the Fermi level (Figure 3.3a). At $(k_x=0, k_y=0)$, the corresponding wave functions of the fourfold degenerate states are localized around four different centers: 1/6, 2/6, 4/6 and 5/6 positions of the enlarged unit cell. The main features of LLs can be realized by discussing one of the fourfold degenerate states, for example, the 1/6-localized Landau states (Figure 3.4b). The quantum number is deter-mined by the normal mode in the B_o sublattice. For the cases of $n^{c, v} \geq 1$ LL, the subenvelope functions of A_o and B_o sublattices are presented in the $(n^{c, v} - 1)$-th and $(n^{c, v})$-th order Hermite polynomials, respectively. They have the follow-ing relationship between conduction and valence states: $A_o^c = A_o^v$ & $B_o^c = -B_o^v$ for the same atoms, and $A_o^v(n^v) \propto B_o^c(n^c = n^v - 1)$ & $B_o^v(n^v) \propto A_o^c(n^c = n^v + 1)$ for the different atoms. It can be deduced with regard to the inter-LL optical tran-sitions that the simple linear relationships account for the specific selection rule $\Delta n = n^c - n^v = \pm 1$, according to the spectral function in Equation 2.37.

For simple hexagonal graphite, the formation of the LSs corresponds to the magnetic quantization of the Dirac cones that are distributed along the K–H line as described by Equation 3.1. The LS energy dispersions strongly depend on k_z, and the relationship between Landau states and wave func-tions at a fixed k_z resembles that of a monolayer graphene. These purely arise from the highly symmetric AA stacking with the same (x,y)-plane projec-tion. At the K point, the conduction and valence LLs are symmetric around $E^{c,v}(n^{c,v} = 0, k_z = 0) = E_D(k_z = 0) \simeq 0.283\alpha_0$, as indicated in Figure 3.5a. The similar LL spectrum is revealed at the H point, while it is centered around $E_D(k_z = \pi / I_z) \simeq -0.279\alpha_0$ (Figure 3.5b).

With the same quantum number, AA-stacked graphite and monolayer gra-phene have the same relationship of two subenvelope functions with respect to the amplitude, spatial symmetry, phase and zero points, as shown in Figures 3.4b and 3.5c. Moreover, the linear relationship between two suben-velope functions remains the same, clearly illustrating that the specific opti-cal selection rule of $\Delta n = \pm 1$ is also applicable to the inter-LS transitions in simple hexagonal graphite. In short, 3D simple hexagonal graphite consist-ing of the same projection graphene layers exhibits the essential 2D quantum phenomena, mainly owing to the Dirac-type energy dispersions.

3.3.2 Landau-Subband Energy Spectra

The LS spectrum in the 1st BZ $(0 \leq k_z \leq \pi / I_z)$ is essential for understanding the magneto-electronic properties of bulk graphites. In the K–H direction, the LSs at a fixed B_0 exhibit a parabolic dispersion with two band-edge states at the two edges of the 1st BZ, that is, K $(k_z = 0)$ and H $(k_z = \pi / I_z)$, as shown for $B_0 = 40$ T in Figure 3.6a. There are no crossings and anti-crossings of LSs, which directly reflects the monotonous dependence of energy bands on wave vectors (the black curve in Figure 3.1a). In particular, the k_z-dependent dispersion of the $n^{c, v} = 0$ LS is consistent with that of the Dirac points along the K–H direction in the absence of external fields, that is, $E^{c, v}(n^{c, v} = 0, k_z) = E_D(k_z)$. A slice of the LS spectrum with respect to a specific k_z can be regarded as a combination of massless Dirac LLs with the zeroth LL given by $E_D(k_z)$. Furthermore, accord-ing to Equation 3.1, the energy width of an LS corresponds to the energy dif-ference between the Dirac points at the zone edges, K and H: that is, $E_D(k_z = 0) - E_D(k_z = \pi / I_z) = (2\alpha_1 + 2\alpha_2) - (2\alpha_1 \cos(\pi) + 2\alpha_2 \cos(2\pi)) = 4\alpha_1 \simeq 1.444$ eV. It should be

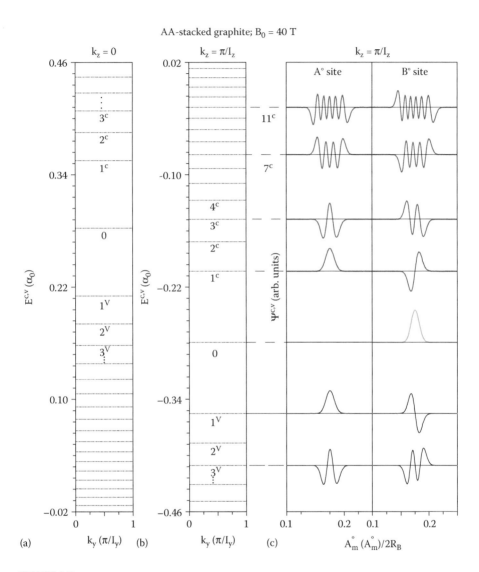

FIGURE 3.5

Landau subbands of simple hexagonal graphite at $B_0 = 40$ T: energy spectrum corresponding to (a) $k_z = 0$ and (b) $k_z = \pi/I_z$; (c) the amplitudes of subenvelope functions of (b).

noted that simple hexagonal graphite retains its semimetallic characteristics in the presence of a magnetic field, implying that free carrier pockets near the K–H edge might cause optical transitions between two valence or conduction LSs (intraband excitations).

On the other hand, with the variation of B_0, the field-dependent energy spectrum displays a form similar to that of monolayer graphene, $E^{c,v}(n^{c,v}, k_z) - E_D(k_z) \propto \sqrt{n^{c,v} B_0}$, while the proportional constant determined by

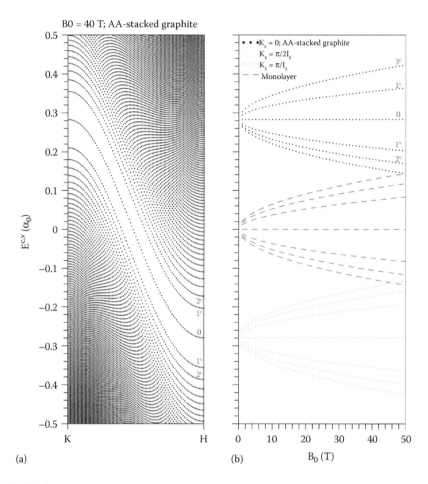

FIGURE 3.6

LS energy spectra of AA-stacked graphite (a) along the KH direction at $B_0 = 40$ T and (b) for the B_0-dependence at various k_zs. Also shown in (b) is the LS energy spectra of monolayer graphene.

the Fermi velocity is weakly dependent on k_z, as shown in Figure 3.6b. In the low-energy approximation, the analytic solution of LS energies is derived by introducing the quantization condition to the Dirac cone of graphites as in Equation 3.1, so that

$$E^{c,v}\left(n^{c,v}, k_z\right) \approx E_D\left(k_z\right) \pm \hbar v_F \sqrt{2eB_0 n^{c,v}/\hbar}.$$ (3.3)

Using these detailed calculations, the four atomic interactions, α_0, α_1, α_2 and α_3, can be expressed in terms of the low-lying LS energies at the K and H points as follows:

$$\alpha_0 = \frac{l_B}{3\sqrt{2}b}\left\{E^c\left(n^c = 1, k_z = 0\right) - E^{c,v}\left(n^{c,v} = 0, k_z = 0\right) + \right.$$

$$\left. + E^c\left(n^c = 1, k_z = \pi/I_z\right) - E^{c,v}\left(n^{c,v} = 0, k_z = \pi/I_z\right)\right\},$$

$$\alpha_1 = \frac{1}{4}\left\{E^{c,v}\left(n^{c,v} = 0, k_z = 0\right) - E^{c,v}\left(n^{c,v} = 0, k_z = \pi/I_z\right)\right\},$$

$$\alpha_2 = \frac{1}{4}\left\{E^{c,v}\left(n^{c,v} = 0, k_z = 0\right) + E^{c,v}\left(n^{c,v} = 0, k_z = \pi/I_z\right)\right\}; \qquad (3.4)$$

$$\alpha_3 = \frac{l_B}{6\sqrt{2}b}\left\{E^{c,v}\left(n^c = 1, k_z = 0\right) - E^{c,v}\left(n^{c,v} = 0, k_z = 0\right)\right.$$

$$\left. - E^c\left(n^c = 1, k_z = \pi/I_z\right) + E^{c,v}\left(n^{c,v} = 0, k_z = \pi/I_z\right)\right\}$$

$l_B = \sqrt{\hbar c / eB_0}$ is the magneto-length related to the effective localization range of LS. Equation 3.4 means that the atomic interactions can be determined by scanning tunneling spectroscopy (STS) and magneto-optical measurements on the LS energies. Based on the band structure, 3D graphite is expected to display massless Dirac-like magneto-optical properties. However, as a result of the strongly dispersed LSs across the Fermi level, the greatly enhanced free carrier pockets near the edges of the first BZ are responsible for the spectral features that are considerably different from the essential quantum phenomena in 2D graphenes.

The magnetically quantized DOS has a lot of special structures, depending on whether it has 1D LSs or 0D LLs. Simple hexagonal graphite exhibits many peaks in the square-root form arising from the quantized LSs with 1D parabolic dispersions, as shown in Figure 3.7a. Each LS contributes two asymmetric peaks corresponding to the band-edge states at the K and H points. For example, at $B_0=40$ T, the $n^{c,v}=0$ ($n^c=1$) LS has two peaks at $E=0.272\alpha_0$ and $-0.272\alpha_0$ ($E=0.35\alpha_0$ and $-0.195\alpha_0$), as indicated by the red (blue) arrows. On the other hand, few-layer graphenes present a plentitude of delta-function-like symmetric peaks due to the dispersionless LLs, for example, the monolayer, bilayer and trilayer systems in Figure 3.7b through d. The initial peak of the zeroth mode (the red arrows) corresponds to the Dirac point (Figure 3.1b through d).

STS is an efficient method of examining the energy spectra of condensed-matter systems. The tunneling differential conductance (dI/dV) is approximately proportional to DOS and directly presents the main features in DOS. The STS measurements have been successfully utilized to identify the diverse electronic properties of graphene-related systems with sp2 bondings, such as few-layer graphenes [69,128–130,235–238], Bernal graphite [64,65], graphene nanoribbons [239,240] and carbon nanotubes [207,208]. Specifically, two low-lying DOS characteristics, a linear E-dependence vanishing at the Dirac point and a $\sqrt{B_0}$ -form LL energy spacing, are confirmed for monolayer

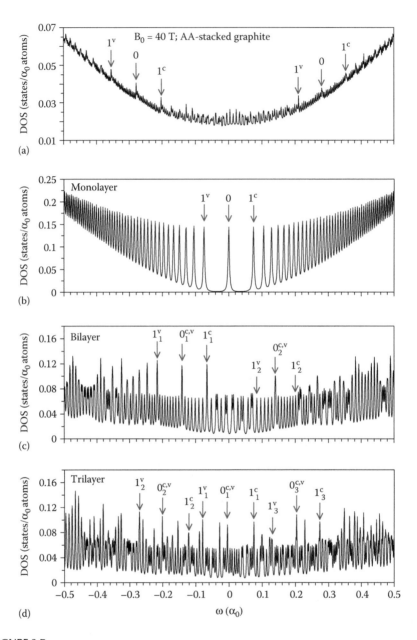

FIGURE 3.7
Magneto-electronic density of states at $B_0 = 40$ T for (a) simple hexagonal graphite, and (b) monolayer, (c) bilayer and (d) trilayer graphenes.

graphene [83,84,235,236]. A sufficiently wide plateau and a lot of square-root LS peaks in AA-stacked graphite require further experimental verifications. The STS measurements on them are useful in the identification of intralayer and interlayer atomic interactions.

3.4 Magneto-Optical Properties

AA-stacked graphite exhibits unique magneto-optical properties as the 1D LSs have sufficiently wide bandwidths and specific energy dispersions. The intraband and interband inter-LS vertical excitations appear in the low-frequency absorption spectra, as clearly indicated in Figure 3.8a through c [57]. The former originate from the valence and conduction LSs across the Fermi level. Only the occupied n^v (n^c) LS to the unoccupied $n^v - 1$ ($n^c + 1$) LS is the effective excitation channel; that is, the optical excitations between the well-behaved Landau wave functions need to satisfy the selection rule of $\Delta n = \pm 1$. The intraband absorption peaks are denoted as ω_{nn-1}^{vv} and ω_{nn+1}^{cc} (Figure 3.8a). They are closely related to the k_z-dependent Fermi-momentum state of each LS ($k_F^{n^{c,v}}$ in Figure 3.9a and c). For example, the ω_{1615}^{vv} peak comes from all the vertical excitations in the range of $k_F^{16^v} \leq k_z \leq k_F^{15^v}$ (Figure 3.9a). ω_{n+1n}^{vv} is close to ω_{nn+1}^{cc}, so their absorption peaks are merged together, for example, those for $n^v \leq 15$ and $n^c \leq 14$ at $B_0 = 40$ T (Figure 3.8a). Such two-channel peaks are observable for $n^{c,v} \leq 6$. As a result of the smaller frequency differences, the other peaks become broad and prominent structures, that is, they behave as multichannel threshold peaks. This composite structure is absent in the layered graphenes and other graphites.

The interband absorption peaks come to exist in the frequency range of $\omega > 0.039\alpha_0$, as clearly shown in Figure 3.8b. They originate from the $(n+1)^v \to n^c$ and $n^v \to (n+1)^c$ vertical excitations, respectively, corresponding to the allowed ranges in $k_F^{(n+1)^v} \leq k_z \leq k_F^{n^c}$ and $k_F^{n^v} \leq k_z \leq k_F^{(n+1)^c}$. $k_F^{n^c} - k_F^{(n+1)^v}$ is almost identical to $k_F^{(n+1)^c} - k_F^{n^v}$, and corresponding excitation frequencies behave similarly (Figure 3.9b). Two kinds of interband channels can create nearly the same absorption spectrum. The effective k_z-ranges are sufficiently wide except for very small quantum numbers, so that the distinct curvature variations of the $(n+1)^v$ and n^c LSs result in two specific absorption frequencies due to the Fermi-momentum states $k_F^{(n+1)^v}$ and $k_F^{n^c}$. Furthermore, such ranges cover the $k_z = \pi/2I_z$ state with the lowest DOS. These are responsible for the existence of many double-peak structures in the cusp form. Such double peaks have non-uniform intensity, and their widths grow with the increasing frequency because of the enlarged range between two associated Fermi-momentum states.

The interband magneto-absorption spectra present unique beating phenomena, as clearly indicated in Figure 3.10a and b.

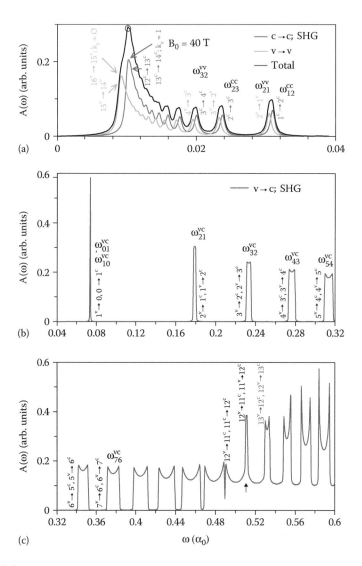

FIGURE 3.8
Magneto-optical absorption spectrum of simple hexagonal graphite in (a) to (c) within different frequency ranges at $B_0 = 40$ T.

The beating oscillations, which include several groups of diversified absorption peaks, are very sensitive to changes in field strength. With the increase of absorption frequency, the widened double-peak structures might overlap with one another. The first group at a lower frequency is composed of isolated double peaks. The second group arises from a combination of two neighboring double peaks, and their composite peak intensity is twice that of the original peaks. With regard to the third group, three neighboring peaks are merged to form a single structure and the intensity of this structure

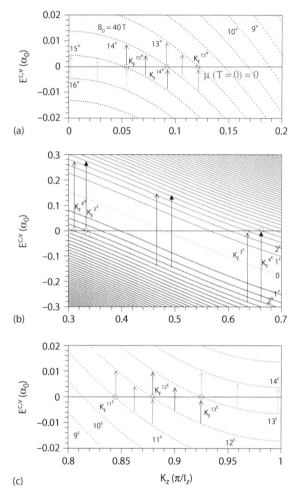

FIGURE 3.9
Vertical optical transitions due to Landau subbands near (a) $k_z = 0$, (b) $k_z = \pi/2I_z$ and (c) $k_z = \pi/I_z$.

is enhanced to almost three times that of the original one. As a result, the spectral intensity is proportional to the number of combined double-peak structures. The unusual association of absorption peaks directly reflects the specific k_z-dependence of each LS parabolic dispersion (Equation 3.3; details in [58]). It should be noted that this is the first time that the beating phenomenon has been predicted in optical properties.

The magneto-absorption peaks of few-layer graphenes and graphite, with the exception of the optical selection rule $\Delta n = \pm 1$, reveal very distinct features. For the former, the dispersionless LLs create delta-function-like symmetric structures with a uniform intensity, as shown in Figure 3.11a.

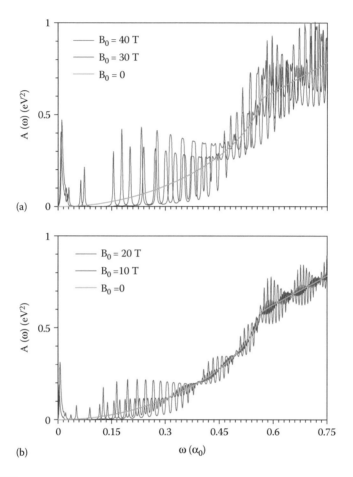

FIGURE 3.10
Beating magneto-absorption spectra of simple hexagonal graphite for various field strengths in (a) and (b).

The multichannel threshold peak is absent. Only one two-channel peak, belonging to the intraband absorption channel, is present in bilayer and trilayer AA stackings (red and blue curves in Figure 3.11a), and it does not have a complete dispersion relation with the B_0-field strength because of the variation of the highest occupied LL (Figure 3.13b and d) [56]. All AA-stacked systems exhibit plenty of interband absorption peaks, but the main differences lie in the peak structures. The monolayer system shows isolated symmetric peaks, while bilayer AA stacking displays pair-peak structures. Furthermore, the N-odd systems correspond to the superposition of the monolayer- and bilayer-like absorption peaks. Some initial $n^v \rightarrow n^c$ excitations are forbidden in $N=2$ and 3 systems, reflecting the Fermi-Dirac distribution of multi-Dirac cones. In addition, the well-behaved beating oscillations are not presented in layered graphenes.

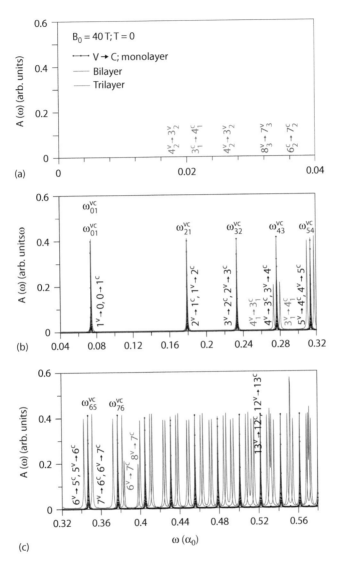

FIGURE 3.11
Magneto-optical absorption spectra of monolayer, bilayer and trilayer graphenes in (a) to (c) within different frequency ranges at $B_0 = 40$ T.

The B_0-dependent absorption frequencies provide important information for the experimental verifications and in understanding the effects of dimensions and stacking configurations. All peak frequencies of AA-stacked graphite, as shown in Figure 3.12a and b, grow with an increase in field strength.

These frequencies present the complete dispersion relations with B_0, in which the field-strength dependence is roughly proportional to except for the multichannel peak (solid circles in Figure 3.12a). The observable intraband

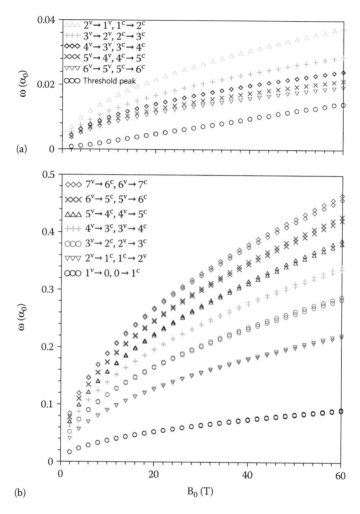

FIGURE 3.12

B_0-dependent absorption frequencies of simple hexagonal graphite corresponding to (a) intraband and (b) interband excitation channels.

excitations cover the multichannel peak and five two-channel peaks. The multichannel threshold frequency does not exhibit a $\sqrt{B_0}$-dependence, as the initial intraband excitation channels dramatically change with field strength. With regard to the interband excitations, there exist two splitting absorption frequencies at sufficiently high magnetic fields. The critical field strength is reduced in the higher-frequency absorption peaks. It is relatively easy to observe the double-peak structures for large ω and B_0. On the other side, the layered graphenes exhibit unique intraband and interband absorption frequencies, as clearly indicated in Figure 3.13a through e. Monolayer graphene has regular $\sqrt{B_0}$-dependence at a low magneto-absorption

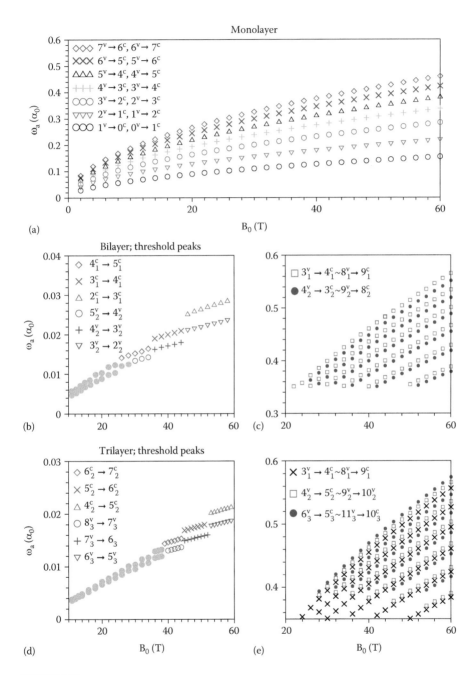

FIGURE 3.13
Magneto-absorption frequencies for (a) monolayer, (b,c) bilayer and (d,e) trilayer graphenes. The discontinuous B_0-dependence of the threshold channel is shown in (b) and (d).

frequency ($\omega < 1$ eV in Figure 3.13a). As to bilayer and trilayer AA stackings, they show discontinuous B_0-dependences in the two-channel intraband peaks (Figure 3.13b and d).

This mainly stems from the fact that the highest occupied LL becomes the smaller $n^{c,v}$ LL with the increase of B_0. Furthermore, their pair-peak interband excitations cannot survive when both n^v and $(n+1)^c$ are occupied or unoccupied, that is, more interband absorption peaks are absent at low field strength (Figure 3.13c and e). The critical differences between three kinds of graphites will be discussed in Chapter 6.

As for magneto-optical measurements, the infrared transmission spectra have identified the $\sqrt{B_0}$-dependent absorption frequencies of the interband LL transitions in monolayer and multilayer graphene [21,23,24,92]. Furthermore, magneto-Raman spectroscopy is utilized to observe the low-frequency LL excitation spectra for AB-stacked graphenes up to five layers [262]. The unique magneto-excitation spectra of simple hexagonal graphite deserve thorough experimental examinations of properties such as the multichannel threshold peak, intraband two-channel peaks, interband double-peak structures and the magneto-optical beating phenomenon. Similar measurements could be made for AA-stacked graphenes to verify the dimension-induced differences in the channel, structure, number, frequency and intensity of magneto-absorption peaks. Such comparisons are useful in illustrating the diversified magnetic quantization of multiple Dirac-cone structures in AA-stacking systems.

4

Bernal Graphite

Bernal graphite, with band profiles of monolayer and bilayer graphenes, is a critical bulk material for a detailed inspection of massless and massive Dirac fermions. Theoretical and experimental research shows that the essential properties of graphite can be described by the quasiparticles at the high symmetry points of the Brillouin zone (first BZ): massless Dirac fermions at the H point and massive Dirac fermions at the K point. In particular, with the dimensional crossover from 3D to 2D, the many exciting properties of few-layer graphenes originate from the interlayer couplings in bulk graphite. The optical excitation channels are only allowed between the respective monolayer-like subbands or bilayer-like subbands, regardless of external fields. The anti-crossings of Landau levels (LLs)/Landau subbands (LSs) and the electron-hole induced twin-peak structures are revealed in both 2D graphene and 3D graphite; however, they are more obvious in graphene with the increase of the layer number. The measured profiles of the B_0-dependent peaks, for example, threshold channels and peak intensity, spacing and frequency, can be used to distinguish the stacking layer, configuration and dimensionality.

4.1 Electronic Structures without External Fields

The band structures of Bernal graphite in the absence of external fields are shown in Figure 4.1a and b. With a slight overlap of conduction and valence subbands, Bernal graphite is classified as a semimetal due to the low-density free carriers. The in-plane energy dispersions considerably depend on the value of the momentum k_z, and contain the characteristics of 2D monolayer and AB-stacked bilayer graphenes at certain special k_zs. In Equation 2.11, $h = 2\cos(k_z I_z)$ indicates the factor of the effective interlayer interactions in Bernal graphite. In the HLA plane ($k_z = \pi/2I_z$ and $h = 2\cos(\pi/2) = 0$), the Hamiltonian matrix can be reduced to a 2×2 matrix of monolayer graphene, because the elements coupling by the nearest-layer interactions are equal to zero and the on-site energy γ_6 can be negligible. It is shown that the occupied valence bands E_v are symmetric to the unoccupied conduction bands E_c around E_F (Figure 4.1b). The low-energy band structure displays a massless

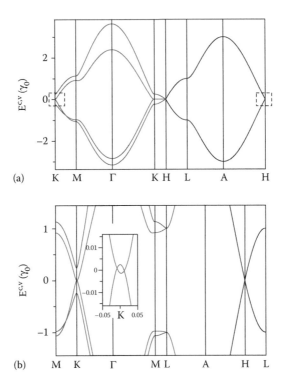

FIGURE 4.1
Band structures for AB-stacked (a) graphite and (b) zoomed-in view at low energies near K and H points.

Dirac-like linear dispersion with the Dirac point located near the H point, while the energy states are double degenerate.

The energy dispersions in the MΓK plane show another graphene property. Substituting the condition $k^z = 0$ and $h = 2\cos(0) = 2$ into the Hamiltonian matrix in Equation 2.11, we find a 4×4 bilayer-like Hamiltonian matrix, while the effective interlayer interactions are twice as large as those of bilayer graphene. The in-plane energy subbands are asymmetric around the Fermi level due to the influence of interlayer atomic interactions, $\gamma_2, \ldots \gamma_6$. In the vicinity of the K point, the low-energy dispersions are characterized by massive Dirac quasiparticles. The wave vectors of the band-edge states are consistent with those of AB-stacked bilayer graphene, that is, the M and K points. However, the effective interlayer atomic interaction $2\gamma_1$ gives rise to double the band-edge state energies $\sim 2\gamma_1$ at the K point in Bernal graphite as compared to bilayer graphene.

Along KH, the strongly anisotropic energy dispersions on k_z are mainly caused by the interlayer interactions. The cosine and flat dispersions along KH (Figure 4.1a) are responsible for the two types of atom chains; one is a straight chain of sublattices coupled by γ_1 along $\hat{z}$, and the other is a zigzag chain of sublattices coupled by γ_4 in the *yz*-plane. In the minimum model, the former and the latter are, respectively, described by $E^{c,v} \simeq \gamma_1 h = \cos(k_z I_z)$ and $E^{c,v} \simeq 0$. When the state grows from K($k_z = 0$) to H($k_z = \pi/2I_z$), the two dispersions gradually get closer and become degenerate at the H point. Also, the in-plane dispersions are bilayer-like, but their behavior transforms into monolayer-like at the H point. That is to say, Bernal graphite exhibits both the massless and massive Dirac fermions in the vicinity of the H and K points, respectively. Angle-resolved photoemission spectroscopy (ARPES) has been used to measure the 3D energy dispersions around the first BZ from the hole pocket at the H point to the electron pocket at the K point [71–75]. Both the massless and massive Dirac fermions are verified in terms of linear and parabolic dispersions, respectively. Furthermore, the measured small-hole pocket at the H point is in agreement with the theoretical model and the quantum oscillation measurements [245]. Remarkably, the Dirac quasiparticles are responsible for special structures in the density of states (DOS) and dominate the optical excitations.

On the other hand, *N*-layer AB-stacked graphenes could exhibit massless and massive Dirac fermions; the band structure resembles a bilayer case or a hybridization of monolayer and bilayer cases, depending on whether the layer number is odd or even. The trilayer graphene displays a hybridization of band structure by monolayer and bilayer graphenes, while the even-layer graphene consists of only pairs of bilayer-like parabolic subbands, as shown in Figure 4.2. Near the K point, the intersection of low-energy subbands indicates that AB-stacked graphenes are gapless 2D semimetals (insets of Figure 4.2a and b). With an increment of the graphene layer, the band structure in cases of even (odd) *N* consists of *N* (*N*−1) pairs of bilayer-like parabolic bands, while it has a particular pair of monolayer-like linear bands near the Fermi level if *N* is odd.

The main characteristics of electronic structures, dominated by the stacking configuration or the interlayer atomic interactions, are directly reflected in the DOS. In Bernal graphite, the DOS mainly originates from the bilayer-like and monolayer-like in-plane dispersions, respectively, corresponding to the K- and H-point band-edge along the k_z dispersions (Figure 4.3). The DOS VHSs (van Hove singularities) marked by black, red and blue colors correspond to the band-edge and saddle-point states in Figures 4.1 and 4.2. A finite DOS at $E = 0$ clearly indicates the semimetallic properties of Bernal graphite. Furthermore, the low-energy intensity smoothly grows with the frequencies, which can be regarded as a superposition of the linear and parabolic dispersions. The former and the latter are, respectively, verified in DOS by the roughly linear and quadratic B_0-dependent tunneling energies [238]. However, a shoulder spreads out near $\pm 2\gamma_1$, which is attributed to the band-edge states of bilayer-like parabolic subbands. Such a structure is reflected by a VHS at $\sim \pm \gamma_1$ in bilayer and trilayer graphenes [69]. With the increasing energies, the DOS exhibits two prominent asymmetric peaks at the middle energies of $E^{c,v} \simeq \pm \gamma_0$. This is a superposition of all saddle points distributed

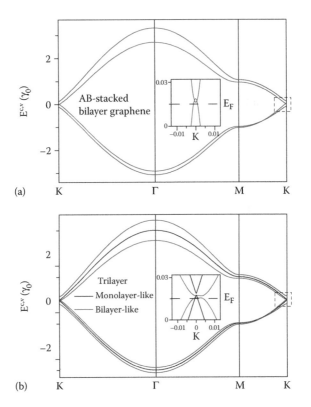

FIGURE 4.2
Band structures of AB-stacked (a) bilayer and (b) trilayer graphenes.

along $M \rightarrow L$ during the band structure transformation from bilayer-like to monolayer-like. In contrast, the trilayer graphene displays three prominent peaks: one comes from the band-edge state of the monolayer-like subband and two from those of bilayer-like subbands. Some of the main features in DOS are verified by scanning tunneling spectroscopy (STS) [69,238] and the measured VHSs could lead to special structures in absorption spectra.

4.2 Optical Properties without External Fields

The absorption is determined by the relationship between the electronic structures (or DOS) and the optical excitation transitions. In Bernal graphite, it demonstrates that $A(\omega)$ is identical for all polarization directions, $\hat{E}$, on the graphene plane, indicating the isotropy of the frequency distribution of the absorption intensity over all frequencies. The optical responses due

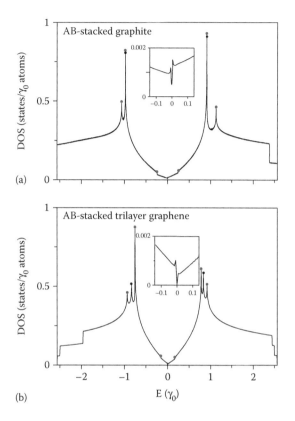

FIGURE 4.3
The DOS of AB-stacked (a) graphite and (b) trilayer graphenes.

to massless and massive Dirac fermions are, respectively, reflected by the optical excitation channels in the vicinity of the H and K points, as shown in Figure 4.4a.

At low energies, one weak shoulder is revealed at $E \simeq 2\gamma_1$ as a result of the excitations between the two low-energy parabolic bands. Moreover, at middle energies, a single sharp peak is accompanied by two shoulders on both its sides. They are responsible for the multi saddle-point channels of all the bilayer-like and monolayer-like band structures as k_z moves from M to L.

Some of the features of the optical spectrum are consistent with the experimental results. Obraztsov et al. [247] study the optical spectra of a polarized beam in cases of different polarizations on the graphene plane and the stacking direction, which are, respectively, indicated as p-polarized and s-polarized. All the cases show a similar behavior between the spectral intensity and polarization of the laser beam, while the photoresponse for the p-polarized excitation beam is relatively strong compared to the s-polarized photoresponse because of the relatively strong energy dispersions on the in-plane

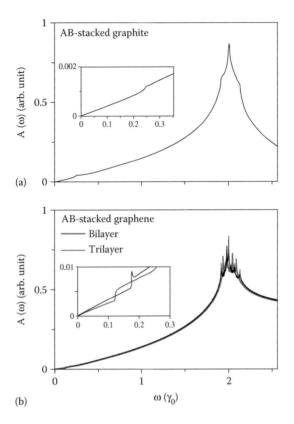

FIGURE 4.4
Optical absorption spectra of Bernal (a) graphite and (b) Graphene.

direction. The results might reflect the relatively strong energy dispersions
for the in-plane direction compared to the out-plane direction.

The optical response of the Dirac quasiparticles is also a dominant contributor for 2D AB-stacked graphenes. It leads to two kinds of special structures:
discontinuities at low frequencies and logarithmic divergences at middle
frequencies. The former and the latter, respectively, come from the vertical
transitions around the K point and those around the M point, as shown by
Figure 4.4b. For the bilayer case, the absorption spectrum exhibits a single
shoulder at $\omega \simeq \gamma_1$ and four peaks at $\omega \simeq 2\gamma_1$. Infrared spectroscopy has shown
a clear picture of the low-energy excitations around the K point [259,260].
Furthermore, the spectral intensity grows with the higher frequency, until
the middle-frequency spectrum when four logarithmic saddle-point peaks
spread around $\omega \simeq 2\gamma_0$. On the other hand, the excitation channels of the
trilayer graphene are only allowed between the respective monolayer-like
subbands or bilayer-like subbands. Infrared spectroscopy has verified the
absorption spectrum, which is a combination of a monolayer and a bilayer
graphene. The aforementioned results indicate that optical spectroscopies

can be used to verify the AB-stacking domains on the surface or the bulk domains of graphite [250].

4.3 Magnetic Quantization

4.3.1 Landau Subbands and Wave Functions

The k_z-dispersed LSs are depicted from the zone boundary point K to H for $B_0 = 40$ T (Figure 4.5a). According to the zero-field band structure, the monolayer-like and bilayer-like signatures are deduced to coexist in Bernal bulk graphite. A series of subenvelope functions distributed among the four

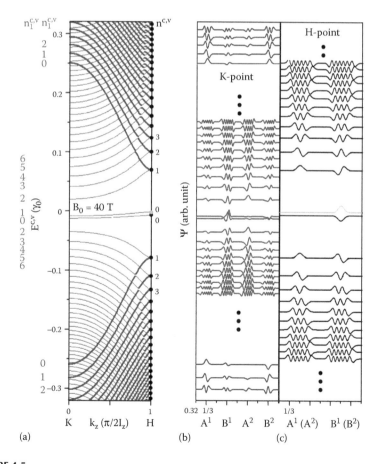

(a)   (b)   (c)

FIGURE 4.5

(a) Landau subbands of Bernal graphite. The subenvelope functions are shown for the (b) K and (c) H points.

constituent sublattices are illustrated in Figure 4.5b and c for the Landau states at K and H points. The LSs can be classified into two groups (blue and red) according to the characteristics of the energy dispersions and the subenvelope functions. In the vicinity of the K point, the two groups are attributed to the magnetic quantization of the respective parabolic subband (blue and red in Figure 4.1). In the 1st BZ, the onset LS energies are consistent with the cosine $E^{c,v} = 2\gamma_1 \cos(k_z I_z)$ and flat $E^{c,v} \sim 0$ dispersions along KH. As k_z changes from K to H, the two groups merge to form a series of double-degenerate monolayer-like Landau states reflecting the zero-interlayer atomic interactions. However, the splitting of the lowest $n^{c,v} = 0$ LSs directly reflects the nonequivalent on-site energies of A and B sublattices. In general, away from the H point, the lift of degeneracy can be mainly attributed to the interlayer atomic interactions γ_1, γ_3 and γ_4 [100]. The energy spacings of LSs are determined by the curvatures of the parabolic subbands. Near the K point, the electron-hole asymmetry of LSs is presented under the influence of the interlayer atomic interactions, while it becomes symmetric for monolayer-like LSs at the H point. Moreover, the K and H points correspond to band edges of LSs, where the Dirac quasiparticles are the dominant contributors to the magneto-optical properties. The effective-mass model (only considering γ_0 and γ_1) can obtain qualitatively consistent calculations for the first few LSs. However, it misses the feature of electron-hole asymmetry, which may be barely observable in STS but has been validated as significant in magneto-reflectance/absorption [89,252,255] and magneto-Raman measurements [253,254].

At the K point, the first group of LSs appears at $E^{c,v} \simeq 0$, and the second group begins at $E^{c,v} \simeq 2\gamma_1$, where the subenvelope functions are associated in a similar way to the case of bilayer graphene. Accordingly, the number of zero points of B_1 and A_1 are employed to define the quantum numbers of the first- and second-group LSs, n_1 and n_2, respectively. That is, the relationship of the two groups is $A_1:A_2:B_1:B_2 = n-1:n-2:n:n-1$ for $n_1 = n > 2$ and $n:n-1:n+1:n$ for $n_2 = n > 1$. As k_z moves to H, with the increasing energy, the first group ascends while the second group descends according to cosine dispersion $2\gamma_1 \cos(k_z I_z)$. Furthermore, the H-point subenvelope functions behave in a monolayer-like way, as depicted by the wave functions in Figure 4.5c.

The profile of monolayer-like (bilayer-like) LSs in the H (K) point can be clearly seen from their energy evolution with the variation of the field strength, as shown in Figure 4.6. Notably, the former is linearly dependent on B_0 and the latter exhibits a square-root B_0-dependence as a consequence of massless and massive Dirac quasiparticles. It is shown in the K-point energy evolution that compared to bilayer graphene, the energy spacings are significantly reduced and the onset energy of the second group of LSs is increased to twice the value of γ_1. This can be simply explained by the minimum model in which the in-plane dispersion at the K point is coupled with a vertical effective hopping energy of $2\gamma_1$ between two neighboring layers. On the other hand, the monolayer-like energy evolution at the H point comes from the fact that the state energies are correlated with only the nearest-neighbor in-plane

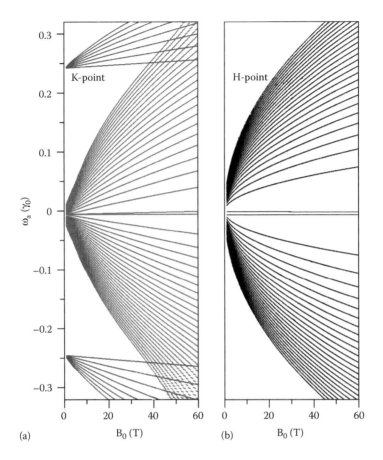

FIGURE 4.6
B_0-dependent energies of (a) K-point and (b) H-point LSs.

hopping γ_0. However, the splitting of the lowest $n^{c,v} = 0$ LSs is revealed only in the case of non-equivalent on-site energies for A and B sublattices.

The DOS also reveals the prominent peaks of both the monolayer-like and bilayer-like signatures, which respectively originate from the vicinity of local extreme values of LSs at the H and K points, as in Figure 4.7a. This implies that the essential properties in Bernal graphite can be regarded as a combination of monolayer and bilayer graphenes, and they also display linear and square-root dependence on B_0. In STS measurements [81], the bilayer-like spectral features are relatively dominant and the valance DOS peaks are stronger than the conduction ones because the corresponding band curvature is relatively small. Furthermore, the energy dependences of the monolayer-like and bilayer-like LLs are also observed in the tunneling spectra of decoupled monolayer, bilayer and trilayer graphenes [84]. It should be noted that the reduced peak spacings, Fermi velocity, effective mass and onset energy of the second group, $2\gamma_1$, are important features to distinguish

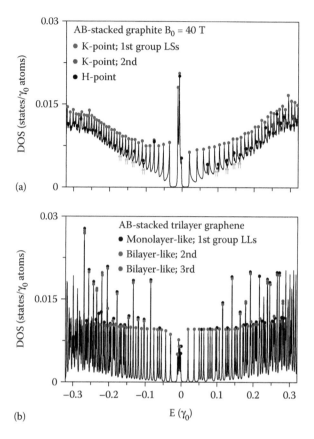

(a)

(b)

FIGURE 4.7
DOS of Bernal graphite at $B_0 = 40$ T. Also shown is the DOS of Bernal trilayer graphene.

the bulk graphite from a bilayer graphene. While the graphene properties have been verified with STS in bulk graphites, there are still unsolved issues regarding the tiny peaks resulting from the bulk properties of LSs between the K and H points.

4.3.2 Anti-Crossings of Landau Subbands

The evolution of the LSs from K to H is responsible for the magnetic quantization that transforms the subbands from parabolic dispersion to linear dispersion, as shown in Figure 4.8. As a result of the anti-crossing of LSs, some tiny peaks appear in couples between monolayer-like and bilayer-like LSs, as indicated by green arrows in Figure 4.7a. Such tiny peaks come from the band extrema of the reversed LSs around $k_z = 0.8$, as shown in Figure 4.8a. In [101], Nakao explained the anti-crossing phenomenon. The author applied the perturbation method to calculate the lift of degeneracy and this led to the conclusion that these level anti-crossings are due to the trigonal warping

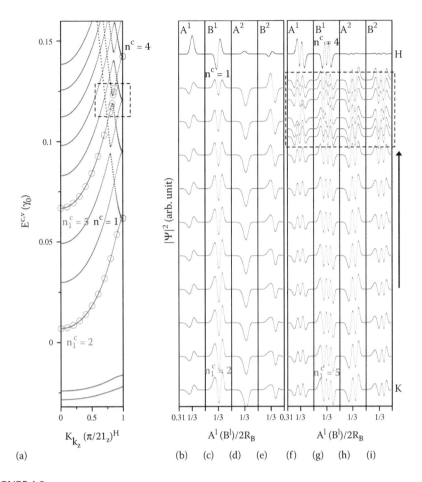

FIGURE 4.8

(a) LS anti-crossing pattern of Bernal graphite; (b)–(i) Evolution of the subenvelope functions along k_z for the low-lying LSs. The dashed rectangular region indicates the hybridized subenvelope functions of the anti-crossing LSs.

effect of γ_3 in the LS spectrum. Also, the event is deduced to appear at the crossover of two LSs that satisfy the condition, $n_1 - n_2 = 3l + 1$, where l is an integer, that is, the two LSs have the same quantum mode for a certain kind of subenvelope function. The opening energy of the anti-crossing LSs is more obvious in the stronger field, a phenomenon also observed for AB- and ABC-stacked few-layer graphenes [31,36,251]. In addition, γ_2 and γ_5 induce band edges for the first few valence subbands near the Fermi level [100]. These VHSs between K–H might be the cause of the unresolved DOS peaks in previous works [81,82].

In the anti-crossing region, the mixture of the LLs caused by the trigonal parameter γ_3 can be realized by the evolution of the subenvelope functions of the LSs. The weight of the amplitude with respect to each sublattice undergoes a significant hybridization along $\hat{k}_z$ for the $n_1^c = 5$ LS that couples with $n_2^c = 1$ LS, as shown in Figure 4.8f through i. The behavior is distinct from the unhybridized LS, for example, the $n_1^c = 2$ LS in Figure 4.8b through e. For the $n_1^c = 2$ LS, as the state moves from K to H, its bilayer-like subenvelope functions gradually transform into monolayer-like ones, that is, the carrier distribution of two layers transfers into one of the two layers. The quantum mode of the dominating sublattice B^1 transforms from two to one. On the other hand, evidence of the state hybridization of $n_1^c = 5$ and $n_2^c = 1$ LSs can be shown by the perturbed behavior of the subenvelope functions around the anti-crossing center, $k_z \sim 0.8(\pi/2I_z)$. Such a state is a multimode state, composed of the main mode $n_1^c = 5$ and the side mode $n_2^c = 1$ LS, as depicted in the dashed rectangle. Nevertheless, at the H point, the wave function of the LS converts to the monolayer-like $n^c = 1$ LS as a result of vanished perturbation. It should be noted that the transition channels of the hybridized LSs might be too weak to observe in optical spectroscopy measurements, but might be observable in STS measurements.

4.4 Magneto-Optical Properties

Bernal graphite, with band profiles of monolayer and bilayer graphenes, is a critical bulk material for a detailed inspection of massless and massive Dirac fermions. The recent surge in interest in 2D graphenes is based on the properties of bulk graphite. The monolayer-like and bilayer-like absorption spectra are predicted to coexist in the bulk spectrum as a result of the excitation channels between intragroup LSs near the H point and between two intragroup and intergroup LSs near the K point. Whether the optical transitions actually take place is subject to the relationship of the initial- and final-state subenvelope functions. According to $A(\omega)$, in Equation 2.37, it is deduced that they must have a quantum-mode difference of one with regard to the same sublattice. The spectral profiles, such as peak intensity, frequency and numbers, can thus be described by the responses of the optical channels of the 1D LSs. In particular, near the band edges of the 1D LSs at the K and H points, charge carriers chiefly accumulate and predominate the optical excitations.

The 1D-LS channels result in square-root divergent peaks in the absorption spectra where two series of peaks, marked by red and black dots, correspond to the massive and massless Dirac fermions, respectively, as shown in Figure 4.9. The K-point associated peaks can mainly be classified into four groups of peaks, resulting from two intergroup and intragroup LS excitations. However, the intergroup ones are obscured due to the broadening peak width and relatively weak coupling of wave functions between the initial state and final state, especially for the region at higher frequencies. Except

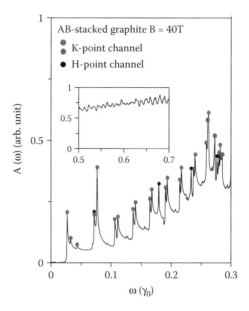

FIGURE 4.9
Magneto-absorption spectra of Bernal graphite. The absorption peaks corresponding to the transitions of massive (massless) Dirac fermions near the K (H) point are marked by red (black) dots.

for the threshold peak that comes from only a single channel, most of the K-point associated peaks come from pair channels, referred to as twin peaks. The splitting energy between the pair is induced by the electron-hole asymmetry; it decreases with increased frequencies or decreased field strengths. In contrast, owing to the subband symmetry at the H point, the degenerate channels give rise to only a single-peak spectrum with a relatively symmetric divergence form. All such peaks can be precisely described by a monolayer graphene with only γ_0. Accordingly, the absorption spectrum of Bernal graphite displays both a bilayer-like twin-peak structure and monolayer-like single-peak structure.

Spectral intensity is mainly determined by two factors: the k_z-dependent band-edge curvatures and the velocity matrix. It turns out that the H-point and K-point excitations make almost the same contribution to the magneto-absorption spectra, despite the former having two times more excitation numbers according to the DOS shown in Figure 4.7. The optical transition rate depends on the expectation value of the velocity matrix, which can be divided into several components of wave function products, each with their own respective integral hopping γ's. However, because γ_0 is at least one order larger than the others, the optical transition rate is simplified as an inner product of the same-layer A and B subenvelope functions of the initial and final states. Based on the k_z-evolution of the subenvelope functions described in Section 4.3.1, the strong–strong combination at H is evidence of the comparable intensity of the Dirac quasiparticles in the H- and K-related spectra.

There are some inconspicuous absorption peaks that come from the band-edge states of the anti-crossing LSs, which satisfy the selection rules of modulo 3, instead of the principle selection rule that characterizes the prominent K-point and H-point peaks in Figure 4.9. One can ascribe the specific selection rules to the hybridization of the anti-crossing LSs, which is determined by interlayer atomic interactions, γ_3 (Section 4.3.2). Therefore, the inconspicuous peaks have relatively weak intensities compared to the K-point and H-point peaks, not only because of the smaller DOS but also because of the smaller velocity matrix. These extra peaks are also studied in graphene systems [137,271], especially severely symmetry-breaking structures, such as AAB- [25] and sliding bilayer graphenes [35]. This indicates that obtaining a comprehensive description of graphene and graphite systems requires all atomic interactions of the Slonczewski–Weiss–McClure (SWM) model for accurate calculations of the velocity matrix and characterization of wave functions.

In graphenes, the spectral peaks are sharper and more distinguishable than those in graphite due to the stronger Landau quantization effect in 2D materials, as shown in Figure 4.10. The excitation channels in bilayer graphene resemble the K-related ones in Bernal graphite, while the absorption peaks are delta-function-like, reflecting the 0D dispersionless LLs as in the 2D systems. This is in contrast to the square-root-like divergent form of the absorption peaks in 3D bulk systems as a result of the 1D dispersive LSs along the out-of-plane direction. Also, the absorption spectrum presents twin-peak structures due to the splitting of dual channels, with a smaller splitting energy of ~10 meV and half-reduced onset energy γ_1, as compared to Bernal graphite. In the case of trilayer AB-stacked graphene, the magneto-absorption spectrum is regarded as a combination of monolayer-like and bilayer-like spectra. The former exhibits a predominantly uniform intensity with single-peak structures similar to the H-point characteristics in graphite. However, the latter displays half-intensity twin peaks corresponding to the splitting of the dual channels in the electron-hole asymmetric LL spectrum. In addition, the absence of intergroup excitations between bilayer-like and monolayer-like LSs is due to the anti-symmetric phase relation in the velocity matrix.

The optical channels at K and H points show two kinds of field evolutions of absorption frequencies, as shown in Figure 4.11. The former type behaves as a linear dependence of B_0 whereas the latter is square-root-like, as depicted by the black and red solid curves in Figure 4.11a. The absorption frequencies related to the H point are identical to those of monolayer graphene and can be well described by a $\sqrt{nB_0}$ relationship (Figure 3.13a). However, compared to the bilayer spectrum (Figure 4.11), the splitting in double peaks is a little

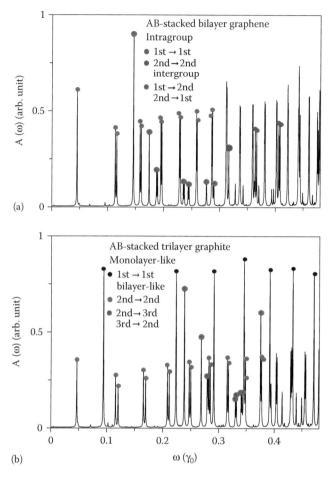

FIGURE 4.10
Magneto-absorption spectra of AB-stacked (a) bilayer and (b) trilayer graphenes.

enhanced for the K-point channels because of the amplification of interlayer atomic interactions in graphite. In short, both AB-stacked graphene and graphite exhibit massless and massive Dirac-fermion properties in the optical absorption spectra with or without magnetic fields. The induced spectral features, such as the frequency dependence, peak width, divergent form and onset energy γ_1, are the signatures that can be used to distinguish between Bernal graphite and graphene. It should be noted that the twin peaks are sensitive to the magnetic field strength. However, the observability in optical spectroscopy depends on how the magnetic field competes with the experiment resolution and ambient temperature. These main features provide very useful information for identifying the stacking configurations and dimensionality of systems from experimental measurements.

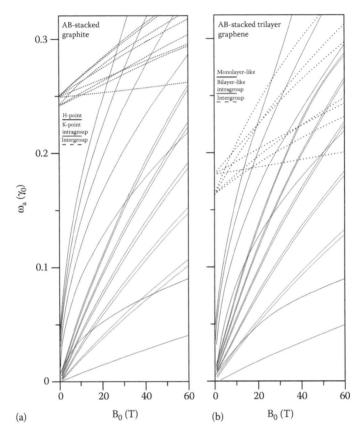

FIGURE 4.11
B_0-dependent absorption frequencies of Bernal (a) graphite and (b) bilayer and trilayer graphenes.

Magneto-optical spectroscopy provides useful insight into the LSs in graphite materials. The optical response of the Dirac quasiparticles is a dominant contributor to the magneto-optical properties. Infrared magneto-transmission studies mainly focus on detailed information about the differences between graphites and multilayer epitaxial graphenes [89–93,246,255,256]. Reflecting bilayer-like parabolic dispersion, a series of absorption peaks of energy scaled as linear-B_0 are present in the spectra and are identified as contributions from massive Dirac fermions in the vicinity of the K point [93]. By satisfying the linear relationship and selection rule $\Delta n = \pm 1$, the effective interlayer interaction $2\gamma_1$ at the K point is obtained in the framework of the minimal model. The deduced value is actually about double that of bilayer graphene. However, reflecting the inherent complexity of the SWM model, there is evidence of the splitting of channels at the K point, which is attributed to electron-hole asymmetry. This is also observed in magneto-reflection [252], magneto-absorption [89,255] and magneto-Raman experiments [253,254]. The experimental results indicate that the full SWM model including additional interlayer atomic interactions well describes the electron-hole LS asymmetry.

The monolayer-like spectra are verified by a series of inter-LS transitions with a characteristic magnetic field frequency dependence $\omega \propto \sqrt{nB_0}$ at the H point [91–93,246,255]. The measured dependence can be used to directly obtain the value of γ_0 in graphite. The full SWM model provides a basic interpretation in such a case to clarify the optical response of the graphene layers and achieve a quantitative agreement between optical experiments and theory. However, there is also evidence of the splitting of degenerate channels at the H point [91,92,246,255]. It is confirmed that the observed splitting is not associated with the electron-hole asymmetry of the Dirac cone. The splitting of the degenerate channels might be attributed to the on-site energy difference between A and B sublattices, spin-orbital coupling, anti-crossing of LSs or parallel magnetic flux. Nevertheless, these results require a more elaborated model and clearer experimental evidence. While the trigonal warping affects the anti-crossings of the LSs in the low-energy region, a new series of absorption peaks obeying new selection rules might possibly be observed with the increased hybridization of the coupled LSs.

Magneto-optical spectroscopies have also been used to study massless and massive Dirac fermions in 2D AB-stacked few-layer graphenes. The electron-hole asymmetry is also reported in bilayer graphene with cyclotron resonance [257], ARPES [79] and infrared spectroscopy [9,259], which is mainly under the influence of γ_2 and the inequivalent environments of the two sublattices. The infrared transmission spectrum of ultrathin graphenes (3–5 layers) indicates that two series of B_0- and $\sqrt{B_0}$-dependent frequencies are observed for the low-lying inter-LL excitations of the massless and massive quasiparticles, respectively [261]. Magneto-Raman spectroscopy has also been used to probe the Dirac-like optical excitations in few-layer graphenes [262,263]. However, further experiments on the higher excitation channels are needed for identifying the energy dependence of the higher LLs away from the Fermi level. The optical experiments can be used to determine the interlayer atomic interactions that dominate the electron-hole asymmetry and LL and LS dispersions. The linear and square-root energy relationships of Dirac-like magneto-channels can be found in few-layer graphenes and graphite, while the differences between the interpreted values of γ's can distinguish the stacking layer, configuration and dimensionality.

5

Rhombohedral Graphite

In an ABC-stacking configuration, 2D graphene and 3D graphite belong to different lattice symmetries. The primitive unit cell of the bulk graphite is a rhombohedron with biparticle lattice symmetry. The essential low-energy properties are described by the 3D anisotropic spiral Dirac cones. However, the dimensional crossover of electronic properties is absent in the ABC stacking. In 2D layered graphene, the surface-localized states appear to be a nontrivial topology of the Dirac spiral in a 3D case. Moreover, the sombrero-shaped subbands, irrelevant to the rhombohedral graphite, are evidence of the 3D to 2D transition. The absorption spectrum is contributed by N^2 channels, while the low-frequency region is mainly dominated by the surface-localized states. Under a magnetic field, each of the N^2 channels displays different B_0-dependence regarding the frequency, intensity and number. Furthermore, the anti-crossing of Landau levels (LLs) leads to discontinuous frequencies and intensities of extra peaks as a function of B_0.

5.1 Electronic Structures without External Fields

In a hexagonal unit cell with $P3$ symmetry [136], the energy dispersions of the rhombohedral graphite are depicted along different symmetric directions, as shown in Figure 5.1. The band structure consisting of three pairs of occupied valence and unoccupied conduction subbands is highly anisotropic and asymmetric around the Fermi level. The in-plane energy bands show linear or parabolic dispersions, whereas they weakly depend on k_z. The valence (conduction) energy bands at K, M and Γ points are the local maximum (minimum), saddle and minimum (maximum) points, respectively. They induce large density of states (DOS) and greatly affect optical excitations. Near the K point, the first pair of subbands, crossing the Fermi level, exhibit linear dispersion without degeneracy. The second pair are double-degenerate parabolic subbands; however, the broken degeneracy along K leads to three nondegenerate parabolic subbands at middle energies $\sim \pm 2B_0$. When the plane is shifted from $k_z = 0$ to $k_z = \pi/3I_z$, similar in-plane dispersions are also revealed in the HLAH plane. However, the pair of linear bands is located at the HA line instead of the HL line. In particular, the energy

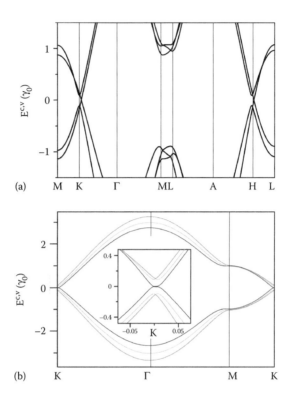

FIGURE 5.1
Band structure of rhombohedral (a) graphite and (b) trilayer graphene.

subband along KH becomes a threefold degeneracy with a very weak k_z-dependent dispersion. Moreover, a small intersection of valence and conduction linear subbands crosses E_F with a very weak dispersion, leading to small free carrier pockets near the K–H region.

Remarkably, one pair of linear subbands always show up at the (k_x, k_y) plane, regardless of the value of k_z. This means that Dirac cones exist in rhombohedral graphite while the Dirac points are spirally distributed along the high-symmetry line K-H. Furthermore, the parabolic subbands are attributed to zone folding [111], because the Hamiltonian (Equations 2.21–2.23) is built in the triple hexagonal unit cell instead of the primitive unit cell, which is a rhombohedron with space group symmetry $R\bar{3}m$.

In the next section, an analytic solution for 3D Dirac cones is calculated in the primitive unit cell along the highly symmetric points using the effective-mass model with only γ_0 and γ_1 [134,265,266]. The trajectory of Dirac-cone movement is a function of k_z. Furthermore, the distortion and anisotropy of the Dirac structures have also been studied separately under the influence of other interlayer atomic interactions [111].

5.2 Anisotropic Dirac Cone Along a Nodal Spiral

The primitive unit cell of a rhombohedral graphite is defined in Figure 5.2a, which is 1/3 of the volume of the hexagonal unit cell. The three primitive unit vectors a_1, a_2, and a_3 are related to the c-axis: $3 I_z \hat{z} = \sum_{i=1}^{3} \mathbf{a}_i$, where $a_1 = a_2 = a_3$, and the angles between two primitive vectors are the same. A rhombohedron with six identical faces is referred to as a primitive cell in ABC-stacked graphite. The Dirac-type dispersion is obtained under a continuum approximation for the low-energy band structure in the vicinity of the H-K-H hexagonal edges specified in Figure 5.2b. The energy dispersion is described as a function of the 3D wave vector with the Brillouin zone (BZ) edge serving as a reference line.

Based on the two sublattices, a full tight-binding Hamiltonian is represented by a 2×2 matrix,

$$H_{ABC} = \begin{Bmatrix} H_1 & H_2 \\ H_2^* & H_1 \end{Bmatrix}, \tag{5.1}$$

where H_1 and H_2 take the form

$$H_1 = 2v_4 \hbar k \cos(\phi + k_z) + 2v_5 \hbar k \cos(\phi - 2k_z),$$

$$H_2 = -v_0 \hbar k \exp(-i\phi) + \beta_1 \exp(-ik_z) + \beta_2 \exp(-i2k_z) \tag{5.2}$$

$$+ v_3 \hbar k \exp[i(\phi - k_z)] + v_5' \hbar k \exp[i(\phi + 2k_z)].$$

The perpendicular wave-vector component k_z is scaled by $1/I_z$ and the variables k, ϕ and v_m' are defined as follows: $k = \sqrt{k_x^2 + k_y^2}$, $\phi = \arctan(k_y/k_x - 7\pi/6)$

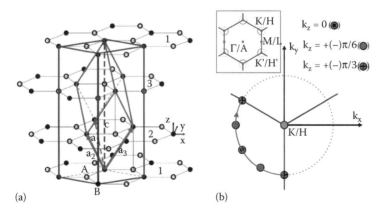

(a)   (b)

FIGURE 5.2
(a) Rhombohedral primitive unit cell (red color) and triple hexagonal unit cell (blue color); (b) (k_x, k_y)-projection of the Dirac-point spiral at $k_z = 0$, $k_z = \pm\pi/6$ and $k_z = \pm\pi/3$.

and $v_m^{()} = 3b|\beta_m^{()}|/2\hbar(m = 0,3,4$ and 5). This chiral Hamiltonian characterizes the inversion symmetry, and its eigenvalues are calculated as

$$E = H_1 \pm |H_2|. \tag{5.3}$$

In Equation 5.1, the off-diagonal elements can be written as $H_2 = f(k_x,k_y,k_z)+\beta_1\exp(ik_z)$ if β_2 and β_5' are neglected, where $|f| \approx [v_0^2 + v_3^2 - 2v_0v_3\cos(2\phi - k_z)]^{1/2}\hbar k$. This indicates the Dirac-cone chirality of massless Dirac fermions, while β_1 induces an offset energy from the hexagonal edge for the Dirac point. On the other hand, the identical diagonal elements lead to gapless Dirac cones, and their linearity in k implies the possibility of cone tilting. Moreover, the zone-folded parabolic bands are absent in the primitive rhombohedral representation.

By ignoring β_2, v_4' and v_5, the coordinate (k_D, ϕ_D) and the energy E_D of the Dirac point are obtained in the case of $H_2=0$; they are, respectively, expressed as follows up to first-order perturbation $\mathcal{O}(v_3/v_0)$:

$$k_D = \beta_1(v_0\hbar)^{-1}\left[1+(v_3/v_0)\cos(3k_z)\right]$$
$$\phi_D = -k_z +(v_3/v_0)\sin(3k_z), \tag{5.4}$$

and

$$E_D\left(k_D(k_z),\phi_D(k_z)\right) = 2\beta_1\left(v_5/v_0 + v_3v_4/v_0^2\right)\cos(3k_z). \tag{5.5}$$

In terms of polar coordinates (q, θ) and the coordinate transformation $q^2=k^2 +k_D^2-2k_Dk\cos(\phi-\phi_D)$, the Dirac-type energy dispersion in Equation 5.3 can be simply expressed as

$$E(q,\theta,k_z) = E_D \pm \epsilon(q,\theta,k_z), \tag{5.6}$$

where

$$\epsilon(q,\theta,k_z) = \left[v_0 - v_3\cos(2\theta - k_z)\pm 2v_4\cos(\theta + k_z)\right]\hbar q. \tag{5.7}$$

This describes the dispersion of the anisotropic Dirac cone; the + and − signs refer to the upper- and lower-half Dirac cones, respectively, and, in particular, the Dirac point E_D displays a spiral dispersion as a function of k_z. In the minimal model with only β_0 and β_1, the Dirac cone is identical to that of monolayer graphene with a Fermi velocity v_0 [134]. The spiral of the Dirac point is described as a function of v_0 in the case of $v_3 = v_4 = 0$, which lies on a cylindrical surface of radius $\beta_1(v_0\hbar)^{-1}$ with the spiral angle ϕ_D in sync with $-k_z$. However, taking into account the interlayer hoppings v_3 and v_4, we find that the spiral becomes non-cylindrical and exhibits k_z-dependent anisotropy.

The former and the latter, respectively, keep and reverse the sign under a phase shift $\theta \rightarrow \theta + \pi$; they, respectively, cause a rotation and tilt of the Dirac cones as a function of k_z. Previous works [111] have demonstrated the aniso-tropic tilt of the Dirac cones that vary in orientation and shape with k_z along the Dirac-point spiral. The Dirac-point spiral across E_F indicates the semime-tallic properties, while the band overlap $\simeq 10$ meV according to Equation 5.5 is one (two) order of magnitude smaller than that of Bernal (simple hexago-nal) graphite. Near $k_z = 0(\pm \pi/3)$, there is an electron (hole) pocket, and at $k_z = \pm 0(\pi/6)$, the free carrier pockets shrink to the Dirac point, that is, the location of the Fermi level $E_D = E_D(\pi/6 = 0)$. This phenomenon is also demonstrated in the hexagonal unit cell (Figure 5.1). Similar analysis can be performed for the H-K-H edge extension, around which the spiral angle synchronizes with k_z and the chirality is reversed.

5.3 Dimensional Crossover

The 3D characteristics of the electronic properties are significant for the dimensional crossover from 2D few-layer graphene to 3D bulk graphite [267]. The ABC-stacked configuration gives rise to different lattice symmetries for the 2D and 3D systems. With the periodic stacking of graphene sheets, the bulk graphite has a biparticle lattice symmetry belonging to the space group $R\bar{3}m$. Its primitive unit cell is a rhombohedron containing two atoms, desig-nated as A and B in Figure 5.2a. In contrast, the N-layer ABC-stacked graphene has two atoms on each layer, that is, a total of $2N$ atoms. Therefore, it is easy to figure out the distinction of the energy bands between both systems. The low-energy electronic properties in the bulk graphite are described by the 3D anisotropic Dirac cones tilted relative to $\hat{k}_z$. On the other hand, the few-layer case is characterized by one pair of partially flat subbands at E_F, which is mainly contributed by the surface-localized states [36,268]. Such appear-ances of subbands, irrelevant to the bulk subbands, indicate the dimen-sional crossover from graphite to few-layer graphene [267]. Nevertheless, the weakly k_z-dependent dispersion across E_F displays a semimetallic behavior for the bulk graphite, making it a candidate system for the observation of 3D QHE [109,134]. Furthermore, there are sombrero-shaped subbands near β_1, which can be used as an interpretation of the dimensional crossover of the 3D case to the 2D limit.

The DOS is very useful for understanding optical properties. Its main characteristics are responsible for the Dirac cones that spiral down as k_z var-ies. In the low-energy region, the DOS intensity increases nonlinearly with the increasing Ω, as shown in Figure 5.3a. At $\omega = 0$, a sharp valley is formed due to the fluctuation of Dirac-point energies within $E_F - 5$ meV $\sim E_F - 5$ meV. Furthermore, a cave structure consisting of a local maximum and local

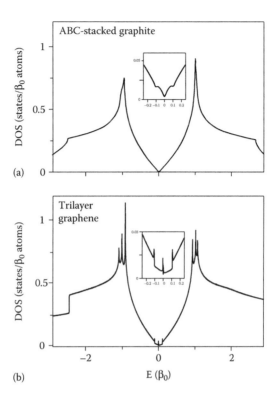

FIGURE 5.3
DOSs of ABC-stacked (a) graphite and (b) trilayer graphene.

minimum is formed near $\omega \simeq (-)0.09\beta_0$ for conduction (valence) DOS [136]. The nonlinear dependence of $D(\omega)$ is attributed to the deformation of the isoenergy surface of the anisotropic Dirac cones, as indicated by the insert of Figure 5.3a. However, when ω exceeds the nonmonotonous structure, $D(\omega)$ becomes linearly dependent on ω as a consequence of the restoration of Dirac cones. In the middle-energy region, the smoothly enhanced DOS is contributed by the parabolic subbands, while the prominent peak near $E^{c,v} \simeq \pm\beta_0$ is a bit broader than that of monolayer graphene as a combination of a series of in-plane saddle-point states in the 3D **k** space.

For ABC-stacked graphene, the dimensional crossover of electronic properties is revealed by 2D divergent structures in the DOS [269]. The DOS is nearly symmetric around the Fermi level, as shown in Figure 5.3b. At low energies, evidence for the partial flat bands near the E_F is revealed by a symmetric broadening peak in the scanning tunneling spectroscopy (STS) measurements [128,129]. Those away from E_F originate from the sombrero-shaped and parabolic subbands. Consequently, the low-energy features in ABC-stacked graphenes are dominated by the surface-localized states. Furthermore, the saddle-point states at middle energies induce three symmetric peaks. These are in contrast to the valley and cusp DOS in the 3D bulk **k** space. The differences between ABC-stacked graphenes and graphites are mainly caused by the different stacking symmetries and the reduction of dimension, which would be reflected in the absorption spectra $A(\omega)$.

5.4 Optical Properties without External Fields

The absorption spectrum of rhombohedral graphite reflects the Dirac-cone energy dispersions, as shown in Figure 5.4a. In general, the low-frequency intensity increases approximately linearly with the increasing ω, as a result of the excitations within the Dirac cones that spiral around the K (K) corner, where the Dirac points fluctuate within a narrow range $E_F - 5$ meV $\sim E_F + 5$ meV [270].

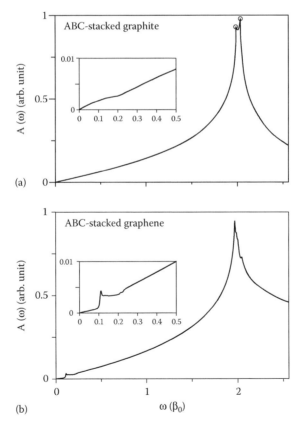

FIGURE 5.4
Absorption spectra of ABC-stacked (a) graphite and (b) trilayer graphene. The insets of (a) and (b) show the zoomed-in view at low frequencies.

However, the interlayer atomic interactions distort the Dirac cones and slightly break the linear dependence of the absorption intensity on the frequency [136]. The small valley at $\sim 0.2\beta_0$ is associated with the transition between the caves in the DOS (Figure 5.4a). As the energy dispersion transforms from linear to parabolic ($\omega \gtrsim 1.0\beta_0$), the spectra deviate from the linear dependence. In the middle-energy region, the enhanced k_z-dependence of the parabolic subbands induces a wider distribution of the spectral structure. There are two separated peaks at $\omega \sim 1.95\,\beta_0$ and $\omega \sim 2.0\,\beta_0$ (black circles); they solely come from the excitations from the BZ edges M and L, respectively. The second peak is higher than the first one because of the higher joint density of states (JDOS) and transition probability.

In an N-layer ABC-stacked graphene, the optical absorption spectrum is richer than that of graphite in both the low- and middle-frequency regions, as a result of the more N N kinds of excitation channels [271,272]. For example, in the trilayer case, the vertical transitions near the K point among different low-lying subbands give rise to feature-rich structures at low energies, including asymmetric peaks and shoulders, as shown in Figure 5.4b. The middle-frequency channels also lead to several obvious peaks associated with the saddle-point states near the M point. It should be noted that the threshold peak, due to the vertical transitions between the surface-localized and sombrero-shaped subbands, is prominent at $\omega \sim \beta_1$ as a dominance of the surface-localized states in the DOS. Optical transmission spectroscopy has been used to verify the optical excitations related to the surface-localized and sombrero-shaped subbands [131,272]. These experimental evidences identify the dimensional crossover from ABC-stacked 3D graphite to 2D graphene.

5.5 Magneto-Electronic Properties

5.5.1 Tight-Binding Model

The LS spectra in rhombohedral graphite are calculated using a diagonalization scheme designed for the Peierls tight-binding Hamiltonian in the representation of the triple hexagonal unit cell [273]. The main characteristics of the spectra exhibit a clearly discernible 3D semimetallic behavior. The LS dispersion is smaller by one or two orders of magnitude than that of Bernal graphite and simple hexagonal graphite, as shown in Figure 5.5a for $\beta_0 = 40$ T. Due to the triple-size enlarged unit cell, the degeneracy of LSs is deduced to be 3×4 for a single LS at a specific k_z point in the hexagonal first Brillouin zone (1st BZ). With the ABC-sacking sequence, its bulk limit has the specific group symmetry $R\bar{3}m$, containing two atoms in a primitive unit cell in comparison to $2N$ atoms for an N-layer 2D case. This results in very distinct magneto-electronic properties between both systems. Accordingly, the LSs

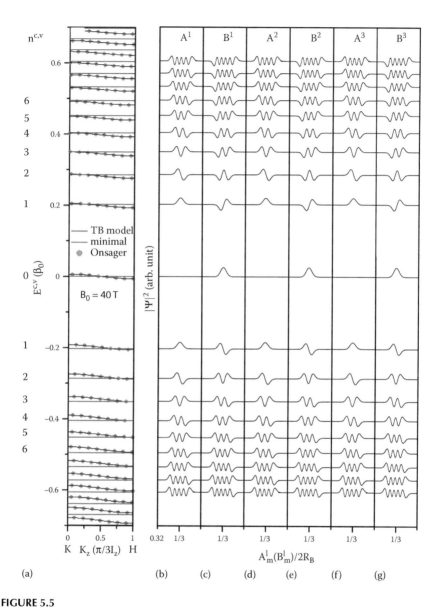

FIGURE 5.5
(a) k_z-dependent Landau subbands at $B_0=40$ T. The black curves, red curves and gray dots, respectively, represent the calculations from the full tight-binding model, the minimal model and the Onsager quantization. (b)–(g) The subenvelope functions at the K point.

(LLs) for the cases of bulk graphite and the N-layer graphene are, respectively, classified as one-group and N-group [31,36,273].

The energy dispersions of LSs actually display a k_z-dependence consistent with the behavior of the zero-field band structure, in which the Dirac-point spiral dispersion in Equation 5.5 corresponds to the $n^{c,v} = 0$

LSs. The behavior of the subenvelope functions provides evidence that also suggests the magnetic quantization of Dirac cones [112]. Their relationship between A and B sublattices is independent on k_z, and the same as that obtained from a comparative diagonalization for monolayer graphene (e.g., at $k_z = \pi/6\, I_z$ in Figure 5.5b through g). The subenvelope function of the $n^{c,v}$ LS consists of quantum modes $n^{c,v}$ and $n^{c,v} - 1$ on sublattices B and A, respectively, where $n^{c,v} - 1 > 0$. In particular, the $n^{c,v} = 0$ LSs are characterized by the same pseudo-spin polarizations on the B sublattice. The reversed pseudo-spin polarization on A is held by the degenerate states of the $= n^{c,v}$ LSs. Within the first-order minimal model, the Dirac-point spirals can be made topologically stable by the chiral symmetry, meaning that the interlayer atomic interactions are not obvious in the diagonalization results [274].

Remarkably, in the minimal model, as a consequence of the magnetic quantization on the Dirac cones, the LS spectrum is dispersionless as a function of k_z, as shown by the red curve in Figure 5.5a. However, the effect of full interlayer interactions reflects the dimensional crossover for the Dirac cone and LS spectra in the rhombohedral graphite. While the properties of Dirac cones are preserved during the variation of B_0, these discernible 3D characteristics, deviated from $\sqrt{n^{c,v}}B_0$-dependence, are presented by B_3 and B_4 in terms of the tilt and distortion of the Dirac cones. In the next section, the Onsager quantization method is used to give analytic energy solutions, and this leads to the identification of the effects of critical atomic interactions on the LSs along a nodal spiral.

5.5.2 Onsager Quantization

The Onsager quantization rule is used to obtain the quantized energies for the Landau states of an isoenergetic surface along the Dirac point spiral [111]. According to the energy dispersion in Equation 5.6, the area $S(\epsilon, k_z)$ enclosed by the contour of energy ϵ is calculated from

$$S(\epsilon, k_z) = \int_0^{2\pi} d\theta \int_0^{\varrho(\epsilon)} \epsilon(q, \theta, k_z) q \cdot dq. \tag{5.8}$$

Using Equation 5.6 and trigonometric substitutions, the integration is approximated as

$$S(\epsilon, k_z) \simeq \pi \epsilon^2 / v_0^2 \hbar^2 \left[1 - (v_3/v_0)^2\right]^{-3/2} \left\{1 + 6(v_4/v_0)^2 \left[1 + 2(v_3/v_0)\cos(3k_z)\right]\right\}. \tag{5.9}$$

The Onsager quantization condition is given by

$$S(\epsilon, k_z) = 2\pi e B_0 \hbar n^{c,v}, \tag{5.10}$$

where the zero phase shift results from the same electron chirality and Berry phase as in monolayer graphene, regardless of the anisotropy of the Dirac cones. By neglecting interlayer atomic interactions, the quantized Landau energies are obtained for an isolated Dirac cone, that is,

$$E^{(0)c,v}\left(n^{c,v}\right) = \pm\hbar v_0\sqrt{2eB_0 n^{c,v} / \hbar}, \qquad (5.11)$$

which turns into

$$E^{(0)c,v}\left(n^{c,v}\right)F\left(k_z\right) \qquad (5.12)$$

for the tilted Dirac cone in rhombohedral graphite, where $F(k_z) = [1 - (v_3/v_0)^2]^{3/4} \times \{1 - 3(v_4/v_0)^2[1 + 2(v_3/v_0)\cos(3k_z)]\}$ is used to describe the dispersion factor as a consequence of the variation of the enclosed area. The energies of LSs are then obtained by superimposing the dispersions (Equation 5.12) on the Dirac point E_D (Equation 5.5), that is,

$$E^{c,v}\left(n^{c,v}, k_z\right) = E_D + E^{(0)c,v}\left(n^{c,v}\right)F\left(k_z\right). \qquad (5.13)$$

According to the lowest LS in Equation 5.13, $E^{c,v}(n^{c,v} = 0, k_z) = E_D \propto \cos(3k_z)$, the Fermi level is determined at the point $k_z = \pi/6$, as the field strength allows the formation of LS bulk gaps, that is, $B_0 \geq [E_D(k_z = 0) - E_D(k_z = \pi/3)]^2/v_0^2 e\hbar \simeq 0.11$ T. Therefore, the renormalized Fermi velocity is given by

$$v_F\left(k_z = \pi/6\right) = v_0\left[1 - \left(v_3/v_0\right)^2\right]^{3/4}\left[1 - 3(v_4/v_0)^2\right]^{1/2}. \qquad (5.14)$$

At $B_0 = 40$ T, the energy dispersions of LSs (Eq. 5.13) are plotted from $k_z = 0 \sim k_z = \pi/3I_z$, as shown in Figure 5.5. The results are consistent with the tight-binding calculations in the energy range, $-\beta_0 \lesssim E^{c,v} \lesssim \beta_0''$.

The B_0-dependent Landau-subband energies display the influence of the interlayer and intralayer atomic interactions, as shown in Figure 5.6. In the minimal model, the LS dispersion and the Fermi velocity are obtained by keeping only v_0 and v_1 in Equations 5.13 and 5.14. The calculated LSs are dispersionless in the 3D momentum space and the Fermi velocity is $v_F = v_0$, as a result of the identical isotropic Dirac cones along a dispersionless Dirac-point spiral. However, the full tight-binding model brings about characteristics beyond the minimal model. Within a Dirac cone, the quantized Landau energies are symmetric around its Dirac point, based on Equation 5.11, while the spiral localization of the Dirac points gives rise to the electron-hole asymmetry. This is interpreted as a consequence of v_3 and v_4, causing a narrow excitation energy range for a single channel, $n^v \rightarrow n^c$, which might be observed in optical experiments. Low-lying LSs are weakly dispersive in contrast to both AA- and AB-stacked graphites. In particular, only the $n^{c,v} = 0$ LS moves across

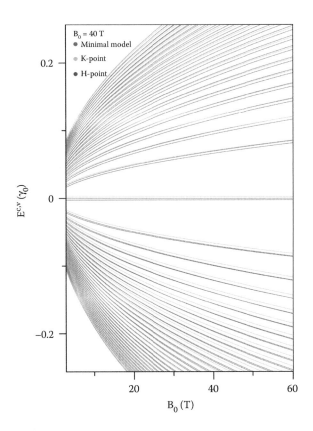

FIGURE 5.6
B_0-dependent Landau-subband energies of rhombohedral graphite.

the Fermi level. The LS spacings are also reduced due to the renormalization of the Fermi velocity. Moreover, the bulk gap can be closed for higher LSs ($n^{c,v} \gtrsim 25$), because of the enhanced k_z dispersion. The LSs calculated from Onsager quantization are consistent with those from numerical diagonalization up to $n^{c,v} = 18 \simeq \pm 730$ meV. However, the inconsistency coming from the continuum approximation is more apparent for higher $n^{c,v}$ values.

The magnetically quantized DOS of rhombohedral graphite is plotted in Figure 5.7a. In the framework of the minimal model, the DOS peaks of rhombohedral graphite are identically reduced to 2D delta-function-like peaks of monolayer graphene, because the Dirac cones are isotropic and circularly distributed at $E_F = 0$. In particular, all the peaks are of equal intensity and symmetric around the Dirac points at $E_F = 0$. Under the influence of β_3 and β_4, the isotropic Dirac cones become tilted, anisotropic and spiraled near the edge of the BZ. The equal-intensity peaks transform into nonequal-intensity double peaks whose widths are determined by the k_z dispersions of the 1D LSs. Each peak has two square-root-divergent forms, corresponding to the band-edge energies at the K and H points (green and blue dots).

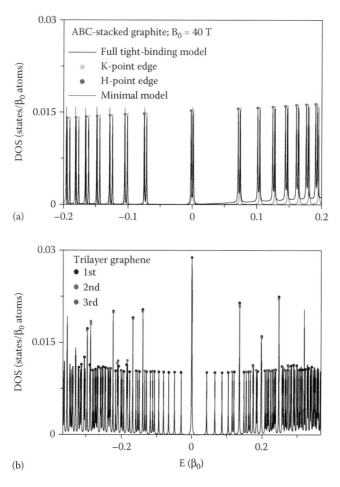

FIGURE 5.7

DOS of ABC-stacked (a) graphite and (b) trilayer graphene under $B_0 = 40$ T. The black and red curves in (a) represent the calculations from the tight-binding model and minimal model, respectively.

The Dirac-point spiral causes the particle-hole asymmetry, which further destroys the anti-symmetric dispersions of LSs in the BZ, leading to different DOS intensities for valence and conduction LSs.

On the contrary, the DOS of ABC-stacked trilayer graphene exhibits three groups of symmetric delta-function-like peaks, which are not regularly sequenced according to the dispersionless LLs of three different subbands, as shown in Figure 5.7b. Different characteristics of the three groups of LLs are clearly shown. The first peak at the Fermi level is composed of three surface-localized LLs; therefore, its intensity is estimated to be around three times that of other peaks. Furthermore, peaks are densely formed for $\omega \gtrsim \beta_1$, which approaches the crossover of the onset energies for the second and third groups. While the low-lying peaks in ABC-stacked few-layer graphenes have been confirmed by STS, the essential differences between ABC-stacked graphites and graphenes need to be further verified.

5.6 Magneto-Optical Properties

The spiral Dirac cones in rhombohedral graphite contribute to a 1D magneto-optical structure, different from monolayer graphene (minimal model) as a result of the tilted anisotropic Dirac cone and the renormalization of the Fermi velocity, as shown in Figure 5.8a for $B_0 = 40$ T. Across the Dirac points, the interband optical transitions between the LSs of n^c and n^v obey a specific selection rule, $n^c - n^v = \pm 1$, at a fixed k_z for the tiled Dirac cones along a nodal spiral [138]. Since $E_D(k_z)$ breaks the anti-symmetric dispersions of $E^{c,\,v}(n^{c,\,v}, k_z)$

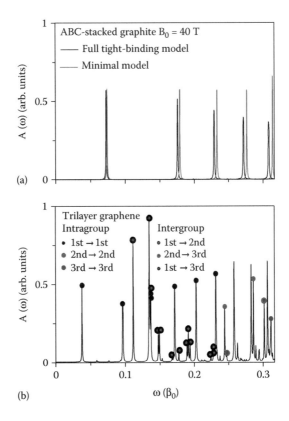

FIGURE 5.8
(a) Absorption spectra of rhombohedral graphite and (b) trilayer ABC-stacked graphene under $B_0 = 40$ T.

along the 1st BZ, the Fermi level is determined at $E_D(k_z = \pi/6)$, exactly across the middle of the zero-mode LSs; therefore, the range of $0 \le k_z < \pi/6$ is unoccupied and that of $\pi/6 \le k_z \le \pi/3$ is occupied. The spectral intensity exhibits a variation during K → H. For a single channel $n^v \to n^c (= n^v \pm 1)$, the peak frequency is calculated from analytic LS solutions in Equation 5.13,

$$\omega_{nn\pm1}^{vc} = \left(\sqrt{2B_0} \hbar v_F / l_B \right)\left(\sqrt{n} + \sqrt{n \pm 1} \right) F\left(k_z = \pi/6 \right), \qquad (5.15)$$

and is accompanied by a frequency distribution between $(\sqrt{2B_0} \hbar v_F / l_B)(\sqrt{n} + \sqrt{n \pm 1}) F(k_z = 0)$ and $(\sqrt{2B_0} \hbar v_F / l_B)(\sqrt{n} + \sqrt{n \pm 1}) F(k_z = \pi/3)$.

Near the BZ edges, K and H, the vertical transitions have only a tiny energy difference and consequently merge to form single peaks; however, within the minimal model, the spectra convert to a uniform-intensity single-peak structure because the LSs are irrelevant to k_z and identical to those of monolayer graphene as described by Equation 5.13 with $E_D = 0$ and $F(k_z) = 1$. Moreover, the peak intensity decreases with the increasing ω due to the fact that the functions $E_D(k_z)$ and $F(k_z)$ vary in terms of interlayer atomic interactions other than β_1. Consequently, under the perturbative k_z-dependent interlayer hoppings β_3, β_4, and so on, the deviation from the massless Dirac-like Landau energies indicates the distortion of the isotropic Dirac cones.

With the dimensional crossover from 3D to 2D, ABC-stacked graphene exhibits magneto-optical properties that are in sharp contrast to those of bulk graphite because of the different lattice symmetries [137,271]. The interband transitions among N groups of conduction and valence LLs contribute to $N \times N$ groups of absorption peaks, each of which displays different B_0-dependence regarding the frequency, intensity and number, as shown in Figure 5.8b [137,271]. In general, the intragroup peaks are relatively stronger than the intergroup ones. The absorption frequencies and intensities increase with the magnetic field strength, and also obey the particular selection rule $\Delta n = \pm 1$. However, the inter-LL excitations for the sombrero-shaped bands give rise to converted frequencies. This abnormal phenomenon is enhanced with the increase of the ABC-stacked graphene layers. Recently, the inter-LL excitation for the partially flat bands and the lowest sombrero-shaped band have been verified by magneto-Raman spectroscopy for a large ABC domain in graphene up to 15 layers [139].

However, abnormal B_0-dependent properties are exhibited by perturbed LLs in small anti-crossing LL regions, as shown in the dashed green ellipse in Figure 5.9b. The corresponding peak intensities and frequencies are discontinuous as a function of B_0 and extra optical selection rules are induced for such LLs with hybridized quantum modes. The larger the layer number N, the more complex the absorption spectrum will be. Furthermore, the spectrum cannot converge to a 3D spiral Dirac absorption spectrum due to the absence of $R\bar{3}m$ symmetry in the 2D limit.

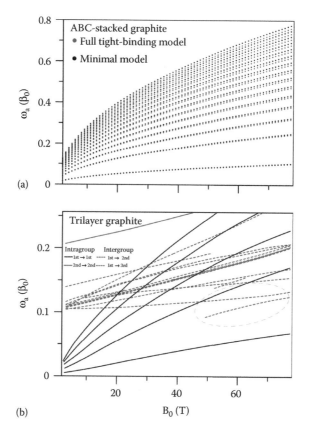

FIGURE 5.9

B_0-dependent absorption frequencies for ABC-stacked (a) graphite and (b) graphene. In (b), the solid and dashed curves represent the intragroup and intergroup absorption frequencies, respectively.

The identification of interband transitions can be made by magneto-absorption, reflection and transmission measurements. The energy width of double peaks, with a separation of ~10 meV, can be resolved; furthermore, it is feasible to probe the Fermi velocity through the measurements of the cyclotron resonance by far-infrared spectroscopy. Moreover, near the Fermi level, rhombohedral graphite has bulk gaps between low-lying LSs in a very weak magnetic field (estimated in Equation 5.13). The achievement of 3D QHE is attainable within the region of bulk gaps; the expected QHE plateaus are different from the experiments reported in ABC-stacked trilayer graphenes [41]. Furthermore, this is in contrast to the cases of AA- and AB-bulk graphite, which have unattainable field strengths that are required to open the bulk gaps to observe a series of 3D QHE plateaus.

6

Quantum Confinement in Carbon Nanotubes and Graphene Nanoribbons

The periodical boundary condition in a cylindrical nanotube surface can quantize the electronic states with the angular momenta ($J^{c,v}$'s) corresponding to the well-behaved standing waves. This leads to a specific optical selection rule when the electric polarization is parallel to the surface [145–148,275,276]. Cooperation with the magnetic field greatly enriches the fundamental properties. Magneto-electronic and optical spectra are very sensitive to changes in the magnitude and direction of the magnetic field and nanotube geometry (radius and chiral angle). The **B**-field can bring about the metal-semiconductor transition, drastically change the 1D energy dispersions, obviously destroy the state degeneracy and induce the coupling of different angular momenta. As a result, there are more **B**-dependent absorption peaks in the square-root asymmetric form. Specifically, magnetic quantization, with high state degeneracy, is absent except for in cases with very large radii and strong magnetic fields.

On the other hand, a finite-size graphene nanoribbon does not have a transverse wave vector, owing to the open boundary condition. The nanoribbon's width, edge structures and external field are responsible for the unusual properties, and it is predicted to present edge-dependent optical selection rules in terms of the subband indices ($J^{c,v}$'s) [168–170]. Based on detailed analyses, these nanoribbons principally come from the peculiar spatial distributions of the edge-dominated standing waves. A perpendicular magnetic field could result in quasi-Landau levels (QLLs), while the magnetic length is longer than the nanoribbon width. Each QLL is composed of partially flat and parabolic dispersions, which dramatically changes the main features of electronic and optical spectra. The magneto-optical selection rule of QLLs is similar to that of monolayer graphene. Graphene nanoribbons are quite different from carbon nanotubes in terms of quantum numbers, wave functions, energy gaps, state degeneracies, selection rules and magnetic field effects, clearly illustrating the critical role of the boundary conditions.

6.1 Magneto-Electronic Properties of Carbon Nanotubes

Each carbon nanotube can be constructed by rolling up a graphene from the origin to the lattice vector $\mathbf{R}x = P\mathbf{a}_1 + Q\mathbf{a}_2$, where $\mathbf{a}_1$ and $\mathbf{a}_2$ are primitive

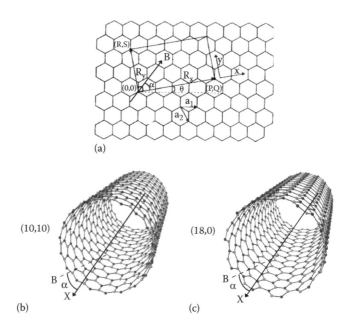

(a)

(10,10)

(b)

(18,0)

(c)

FIGURE 6.1
Geometric structures for (a) graphene and carbon nanotubes; (b) (10,10) and (18,0) nanotubes.
Also shown in (b) is the relation between nanotube axis and magnetic field.

lattice vectors of a 2D sheet (Figure 6.1a). A (P,Q) nanotube has a chiral angle
of $\theta = tan^{-1}[\sqrt{3}\,Q/(2P+Q)]$ and radius of $r = b\sqrt{3(P^2+PQ+Q^2)}/2\pi$. (P,P)
and $(P,0)$, respectively, correspond to nonchiral armchair and zigzag sys-
tems ($\theta = 30°$ and $0°$). The number of carbon atoms in a primitive unit cell is
$N_c = 4\sqrt{(P^2+PQ+Q^2)(R^2+RS+S^2)/3}$, where (R,S) correspond to the primitive lattice vector along the nanotube axis.

The misorientation of $2p_z$ orbitals (the curvature effect) on the cylindrical surface leads to the change in the hopping integral, that is, $\gamma_i = V_{pp\pi}\cos($
$\Phi_i)+4(V_{pp\pi}-V_{pp\sigma})\,[r/b\,\sin^2(\Phi_i/2)]^2$, where $i(=1, 2, 3)$ corresponds to the three
nearest neighbors, and Φ_i ($\Phi_1 = -b\,\cos(\pi/6-\theta)/r$, $\Phi_2 = b\,\cos(\pi/6+\theta)/r$; $\Phi_3 = -b$
$\cos(\pi/2-\theta)/r$) represents the arc angle between the two nearest-neighbor
atoms. The Slater–Koster parameters $V_{pp\pi}(=-2.66$ eV$)$ and $(=6.38$ eV$)$, respectively, represent the π and σ bondings between two $2p$ orbitals in a graphene
plane [281]. The Hamiltonian matrix will be built for any magnetic-field
direction. α is the angle between the magnetic field and axis. As for a zero
or parallel magnetic field, a 2×2 Hamiltonian, accompanied by $J^{c,v}=1,2,3,$
and so on, and $N_c/2$, is sufficient for calculating the essential properties [280].
On the other hand, the different J's would couple one another as the magnetic field deviates from the nanotube axis. The perpendicular component
$B_\perp = B_0 = \sin$ leads to the J coupling and the total carbon atoms in a primitive
cell are included in the tight-binding calculations (details in [147,279]).

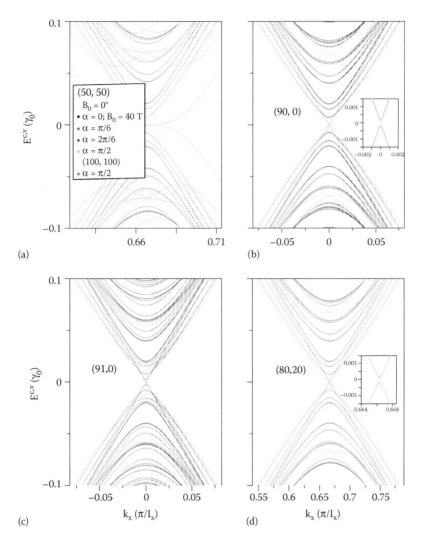

FIGURE 6.2
Band structures of (a) (50,50), (b) (90,0), (c) (91,0) and (d) (80,20) carbon nanotubes at various magnetic fields and with $B_0 = 0$ and 40 T. Also shown in (a) is the band structure of a (100,100) nanotube at $\alpha = \pi/2$ (orange curves), and in the insets of (b) and (d) are narrow energy gaps.

Each carbon nanotube has many 1D energy subbands at zero field, which are defined by angular momenta of $J^{c,v}$'s, as shown in Figure 6.2. For a (P, P) armchair nanotube, a pair of linear valence and conduction bands, with $J^{c,v} = P$, intersects at E_F, for example, those of $J^{c,v} = 50$ in a (50,50) nanotube (gray dotted curves in Figure 6.2a). There, they create a finite density of states (DOS) (Figure 6.4a) and thus behave as a 1D metal. The fact that the metallic behavior is not affected by the curvature effect can be identified from the periodical boundary condition and the specific changes of the nearest-neighbor

hopping integrals [141]. That is, the Fermi-momentum states in armchair carbon nanotubes are sampled from the Dirac points of a graphitic sheet. In
addition, the misorientation of $2p_z$ orbitals only leads to a slight redshift in
the Fermi momentum (k_F), compared with 2/3 (in unit of $\pi/I_y;I_x$, the periodical distance along the nanotube axis). The higher/deeper energy subbands
are doubly degenerate, as they correspond to $J^{c,v}$ and $N_c/2 - J^{c,v}$ simultaneously. They present parabolic dispersions near the band-edge state of $k_x = k_F$.
    The radius and chiral angle can determine the energy gap and state degen-
eracy. (P, Q) carbon nanotubes, respectively, belong to narrow- and moderate-gap semiconductors, and are characterized by $(2P + Q = 3I$ & $P \neq Q)$ and
$2P + Q \neq 3I$. The former and the latter, with energy gaps inversely proportional
to radius and the square of radius, arise from the periodical boundary condition and the curvature effect, respectively [141,280]. For example, the narrow-gap (90,0) and (80,20) nanotubes and the moderate-gap (91,0) nanotube,
respectively, have $E_g \sim 0.0005\,\alpha_0$ (Figures 6.2b and d) and $0.03\,\alpha_0$ (Figure 6.2c).
In general, the low-lying valence and conduction bands in semiconducting
nanotubes possess parabolic dispersions with double degeneracy.
    State degeneracy, band gap and energy dispersions strongly depend on
the direction and strength of a uniform magnetic field. When **B** is parallel
to the nanotube axis, the angular momentum becomes $J^{c,v} + \phi/\phi_0$ (magnetic
flux $\phi = \pi r^2 B_0$). The gapless linear bands of armchair nanotubes change into
separate parabolic bands (black dotted curves in Figure 6.2a), that is, the
metal-semiconductor transition occurs in the presence of ϕ. Furthermore, the
magnetic flux destroys the double degeneracy in the higher/deeper energy
subbands; that is, two subbands characterized by the angular momenta of
$J^{c,v}$ and $N_c/2 - J^{c,v}$ are not identical to each other because of the magnetic
splitting effect. Such effects are relatively prominent when the direction of
the magnetic field approaches the nanotube axis. On the other hand, a nonparallel magnetic field will result in the coupling of different $J^{c,v}$'s, which
will be stronger at large r, B_0 and α values. Specifically, a perpendicular
magnetic field can break the state degeneracy, in which the splitting subbands are symmetric around $k_x = 2/3$ (green dotted curves). An armchair
nanotube keeps the metallic behavior at $\alpha = \pi/2$. Whether the intersecting
linear energy bands become gapless parabolic ones depends on radius and
field strength. For example, (50,50) and (100,100) nanotubes, respectively,
exhibit linear and parabolic dispersions (green and orange dotted curves).
Moreover, the **B**-induced effects on band structures are clearly revealed in
non-armchair carbon nanotubes (Figure 6.2b through d).
    Diverse relations exist between energy gap and magnetic field. Under a
parallel magnetic field, energy gaps present a linear and nonmonotonous
relation with a period of ϕ_0, as clearly indicated in Figure 6.3. The E_g of an
armchair nanotube grows with the increase of ϕ, reaches a maximum value
at $\phi = \phi_0/2$, and then recovers to zero at ϕ_0 (black curve in Figure 6.3a). The
metal-semiconductor is also revealed in the narrow- and middle-gap car-
bon nanotubes (black curves in Figure 6.3b and c) near small ϕ and $\phi_0/3$

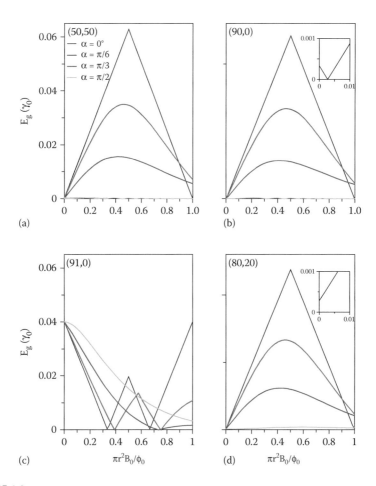

FIGURE 6.3
Magnetic-flux-dependent energy gaps for (a) (50,50), (b) (90,0), (c) (91,0) and (d) (80,20) carbon nanotubes at various magnetic-field directions.

(or ϕ_0 and $2\phi_0/3$). In addition, they might be absent in part of narrow-gap systems (inset in Figure 6.3d). Obviously, the periodical Aharonov–Bohm effect is presented in energy gaps of any carbon nanotube. It would be very difficult to observe this effect in the presence of a dominating perpendicular magnetic field, for example, E_g's at large α's. There are no simple relations between energy gaps and ϕ under nonparallel fields. As for the middle-gap nanotubes (Figure 6.3c), the larger the α value, the higher the B_0-field strength of the semiconductor-metal transition. Specifically, at $\alpha=90°$, both armchair and narrow-gap nanotubes keep the gapless feature (green curves in Figure 6.3b and d), that is, the metal-semiconductor transition does not occur during the variation of field strength. The predicted strong dependence of energy gap on the strength and direction of the magnetic field and

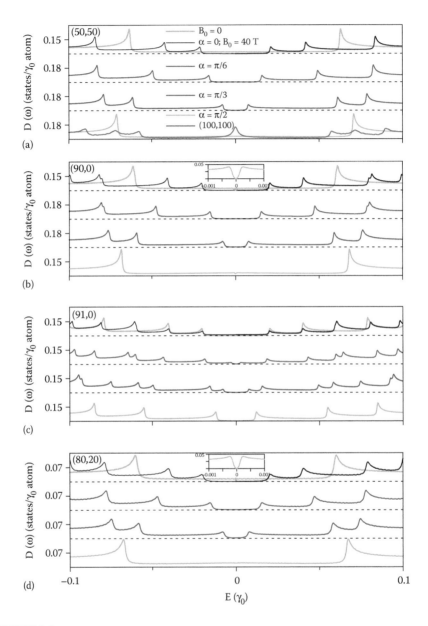

FIGURE 6.4

Density of states for (a) (50,50), (b) (90,0), (c) (91,0) and (d) (80,20) carbon nanotubes at various magnetic-field directions.

nanotube geometry could be further verified by scanning tunneling spectroscopy (STS) measurements.

The special structures in DOS are greatly enriched by nanotube geometry and the magnetic-field direction, as shown in Figure 6.4. The 1D band-edge states in parabolic and linear dispersions, respectively, create square-root asymmetric peaks and plateaus. All VHSs (van Hove singularities) belong to the former except for that near $E=0$ in metallic nanotubes at zero field (gray solid curve in Figure 6.4a). For an armchair nanotube, a parallel magnetic field makes the plateau change into a pair of asymmetric peaks with an energy gap (black curve), and the number of intensity-reduced asymmetric peaks at other energies doubles. The energy spacing between two neighboring peaks declines as α grows (red, blue and green curves). Furthermore, they are merged together under a perpendicular field (green [orange] curve), in which there is a plateau structure (a symmetric peak) near $E=0$ because of a pair of gapless linear (parabolic) bands (Figure 6.2a). Compared with metallic armchair nanotubes, the narrow-gap systems, as shown in Figure 6.4b and d, present almost the same structures except for a pair of very close asymmetric peaks near the Fermi level at zero field (insets). However, there are more peak structures in the moderate-gap nanotubes (Figure 6.4c). On the experimental side, the zero-field DOS characteristics, including the asymmetric peaks, energy spacings, metallic behaviors [207,208] and middle and narrow gaps [142], have been verified by STS measurements. The rich peak structures and the metal-semiconductor transitions due to a uniform magnetic field require further experimental verifications in DOS.

6.2 Magneto-Optical Spectra of Carbon Nanotubes

The standing waves on a cylindrical surface play a critical role in determining the available optical excitation channels. At zero field, they present well-behaved spatial distributions, with the forms of sine/cosine functions closely related to the angular momenta, as revealed in Figure 6.5a through c. For an armchair nanotube, the linear energy bands possess a uniform distribution along the azimuthal direction (gray lines in Figure 6.5a), and the higher-energy bands correspond to the normal oscillations of one and two wavelengths (Figure 6.5b and c). Such features are independent of k_xs and remain the same in the presence of a uniform magnetic field, leading to the identical optical selection rule at zero and parallel magnetic fields. In addition, these two factors only create rigid shifts in azimuthal distributions. As to the nonparallel magnetic fields, normal standing waves are changed into distorted ones, perpendicular ones in particular (green curves in Figure 6.5a through c). However, the latter could be regarded as a superposition of the J-decoupled normal modes, as a perpendicular field creates the coupling of distinct angular momenta. It is relatively easy to observe the coupling effect in larger nanotubes. For example, an armchair (100,100) nanotube exhibits highly distorted standing waves with more zero points (orange curves in Figure 6.5e), corresponding to the gapless parabolic energy bands near E_F and the oscillatory ones (Figure 6.2a).

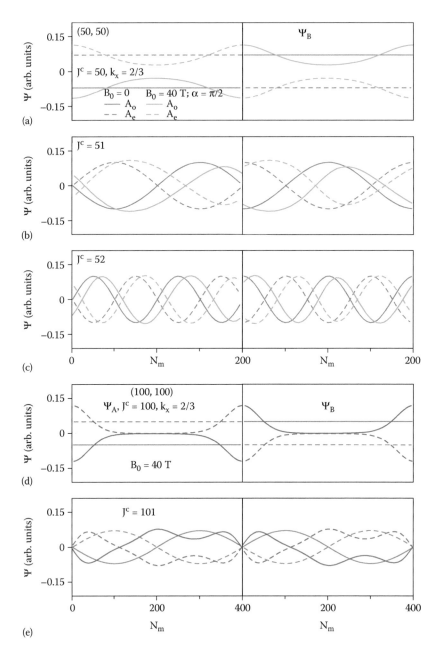

FIGURE 6.5
Subenvelope functions of the (50,50) nanotube at $k_x = 2/3$ for (a) $J^c = 50$, (b) $J^c = 51$ and (c) $J^c = 52$. $B_0 = 0$ and ($B_0 = 40$ T, $\alpha = \pi/2$) are shown by the black and red curves, respectively. Also shown are those of the (100,100) nanotube for (d) $J^c = 100$ and (e) $J^c = 101$.

The electric polarization is assumed to lie on the nanotube surface (parallel to the nanotube axis), as considered in layered graphites (Equations 2.37 and 2.38). All the carbon nanotubes only exhibit asymmetric absorption peaks in the square-root form, as shown in Figure 6.6. The available excitation channels arise from the occupied valence subbands and the unoccupied conduction ones, and have the same angular momentum. The selection rule

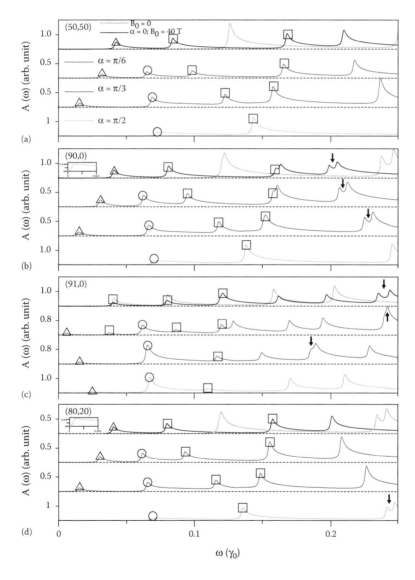

FIGURE 6.6
Optical absorption spectra of (a) (50,50), (b) (90,0), (c) (91,0) and (d) (80,20) carbon nanotubes under zero-field and various magnetic-field directions.

of $\Delta J = J^v - J^c = 0$ is determined by the J-decoupled standing waves. This rule could be further applied to any magnetic field, even with the coupling of J's, by using the superposition of distinct components. At zero field (gray curve in Figure 6.6a), an armchair nanotube does not have the threshold absorption spectrum as the vanishing velocity matrix elements (Equation 2.38) prevent interband excitations due to a pair of linearly intersecting energy bands. The featureless optical spectrum is also revealed in monolayer graphene (Figure 3.3). The absorption peaks are closely related to the band-edge states of the parabolic valence and conduction bands, that is, the absorption frequency is their energy spacing. In effect, the weaker the energy dispersion, the stronger the asymmetric peak.

The number, frequency and intensity of absorption peaks are very sensitive to the magnetic field. The threshold absorption peak of an armchair system is generated by a parallel magnetic field (triangle related to the black curve in Figure 6.6a). The number of other absorption peaks becomes double (two rectangles) and their intensities grow weak, directly reflecting the splitting of the J-dependent degeneracy (Figure 6.2a). All the absorption peaks agree with the $\Delta J = 0$ rule. This rule is also suitable for most peak structures, even when a magnetic field gradually deviates from the nanotube axis. The first peak declines quickly and two neighboring peaks approach each other (red and blue curves): the former disappears and the latter changes into a single peak under a perpendicular magnetic field (green curve). It should be noted that the absence of a threshold peak at $\alpha = \pi/2$ is independent of the radius of the armchair nanotube and field strength. On the other hand, there exists an extra low-frequency absorption peak (circle in red, blue and green curves), mainly owing to the vertical excitations of the first (second) valence band and the second (first) conduction band (Figure 6.2a). This peak does not satisfy the selection rule of $\Delta J = 0$, but it could be identified from the coupling of the neighboring angular momenta.

The main features of absorption peaks strongly depend on the geometric structures of carbon nanotubes. The narrow-gap systems exhibit a zero-field threshold peak at a very low frequency (insets in Figure 6.6b and d). For a non-perpendicular B_0-field, they might have merged double-peak structures (arrows in Figure 6.6b), reflecting the small energy spacings between two neighboring parabolic subbands with high DOSs (Figure 3.3b). In addition, an extra prominent magneto-absorption peak of $\Delta J \neq 0$ is revealed in all carbon nanotubes at $\alpha \neq 0°$ (circles in Figure 6.6) because of the coupling of distinct angular momenta. As for the moderate-gap nanotubes, there are more prominent absorption peaks (Figure 6.6c) arising from the rich, low-lying parabolic subbands (Figures 6.2c and 6.3c).

The dependence of absorption frequencies on the magnetic-field direction is important in understanding the characteristics of prominent peaks, mainly owing to the composite effects arising from the splitting and coupling of angular momenta. For example, within the frequency range of $\omega \leq 0.25\gamma_0$, all the magneto-absorption peak frequencies monotonously grow or decline as the magnetic field gradually deviates from the nano-tube axis, as clearly shown in Figure 6.7. This is related to the band-edge state energies of the split subbands. The transition intensities will undergo a drastic change during the variation of α, so that absorption peaks might disappear or come to exist. The threshold peak (triangle) becomes absent in the metallic (narrow-gap) carbon nanotubes under a large α, for example, $\alpha = 86°$ for the (50,50) nanotube in Figure 6.7a. However, it could survive in the middle-gap systems for any angles (Figure 6.7b). Its disappearance and existence, respectively, correspond to the intersecting linear bands and the gapped parabolic bands (green curves in Figure 6.2a and c). An extra peak (circle), reflecting the significant coupling effect, is revealed at a large enough $\alpha(\sim 3.6°–5.4°)$. Most magneto-absorption peaks (rectangles) present splitting behavior. Whether they are merged together or become absent

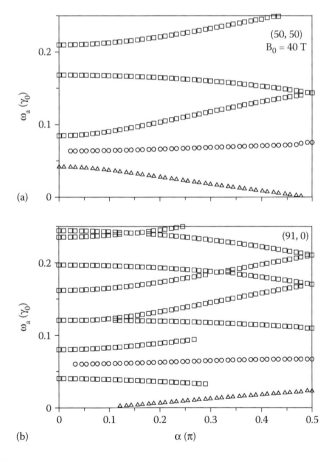

FIGURE 6.7
Optical absorption frequencies of (a) (50,50) and (b) (91,0) carbon nanotubes under various magnetic-field directions at $B_0 = 40$ T.

depends on α. Up to now, the splitting peaks due to a parallel magnetic field have been confirmed by optical measurements [149]. The predicted strong dependence on the magnetic-field direction and nanotube geometry is worthy of further examination.

6.3 Magneto-Electronic Properties of Graphene Nanoribbons

Two typical kinds of achiral graphene nanoribbons, with the hexagons normally arranged along the edge structure, are chosen for a model study. Zigzag and armchair graphene nanoribbons, respectively, correspond to armchair and zigzag carbon nanotubes. Their widths can be characterized by the number of zigzag and dimer lines (N_y) along the transverse y-direction, respectively (Figure 6.8a and b). The low-energy band structures are evaluated from the $2N_y$ tight-binding functions of $2p_z$ orbitals. Such functions are combined with the Peierls phases to explore the magneto-electronic and optical properties. For a zigzag ribbon, the Peierls tight-binding Hamiltonian matrix can be expressed as

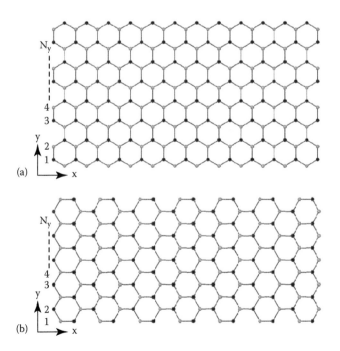

FIGURE 6.8
Geometric structures of (a) zigzag and (b) armchair graphene nanoribbons. N_y is the number of zigzag or dimer lines.

$$H_{ij} = \begin{cases} 2\gamma_0 \cos\left(k_x \dfrac{\sqrt{3}b}{2} + \Delta G_1\right) & \text{for} \quad j=i+1, \quad j \quad \text{is even,} \\ \gamma_0 & \text{for} \quad j=i+1, \quad j \quad \text{is odd,} \\ 0 & \text{others,} \end{cases} \quad (6.1)$$

where the Peierls phase difference $\Delta G_1 = -\pi\phi(j/2 - [N_y+1]/2)$.

As to an armchair ribbon, the Hamiltonian matrix is given by

$$H_{ij} = \begin{cases} \gamma_0 \exp i\left(-k_x b + \Delta G_1\right) & \text{for} \quad j=i+1, \quad j \quad \text{is even,} \\ \gamma_0 \exp i\left(k_x b/2 + \Delta G_2\right) & \text{for} \quad j=i+3, \quad j \quad \text{is even,} \\ \gamma_0 \exp i\left(-k_x b/2 + \Delta G_3\right) & \text{for} \quad j=i+1, \quad j \quad \text{is odd,} \\ 0 & \text{others,} \end{cases}$$

where $\Delta G_1 = \pi\phi(j/2 - [N_y+1]/2)$, $\Delta G_2 = -\pi\phi/2(j/2 - 1 - [N_y]/2)$ and $\Delta G_3 = \pi\phi /2[(j+1/2)-1-[N_y]/2]$.

The dimension of the Hamiltonian stays the same, even in the presence of a perpendicular magnetic field, while it is largely enhanced for layered graphenes [36] or graphites (Chapter 2). The strong competition between the finite-size confinement and magnetic quantization will greatly diversify the essential properties.

The finite-size effect directly induces plenty of 1D energy subbands, as shown in Figure 6.9. Each subband does not have a transverse quantum number under the open boundary condition; furthermore, the $J^{c,v}$ index only represents the arrangement order measured from the Fermi level. This is one of the most important differences between graphene nanoribbons and nanotubes. Electronic structures strongly depend on the edge structures. An $N_y = 100$ zigzag nanoribbon exhibits band-edge states near $k_x = 2/3$ (Figure 6.9a), similar to those in a $(50, 50)$ armchair nanotube (Figure 6.2a). All the low-lying energy subbands possess parabolic dispersions except for the first pair with partially flat ones at larger k_x's. The latter belong to the edge-localized states (discussed later). On the other hand, the band-edge states are situated at $k_x = 0$ for an armchair nanoribbon, for example, $N_y = 180$ in Figure 6.9e. Each energy subband has no double degeneracy except for the spin degree of freedom. The energy gap, which is determined by the parabolic valence and conduction subbands of $J^{c,v} = 1$, declines with the increase of ribbon width. The width-dependent energy gaps have been confirmed by STS measurements [164–167].

The energy dispersions are dramatically changed by magnetic quantization. The energy spacings of a zigzag nanoribbon become non-uniform during the variation of B_0, as shown for $N_y = 100$ in Figure 6.9b at $B_0 = 40$. When the field strength is high enough, the lower-energy 1D parabolic subbands will evolve into composite subbands (QLLs) with parabolic and

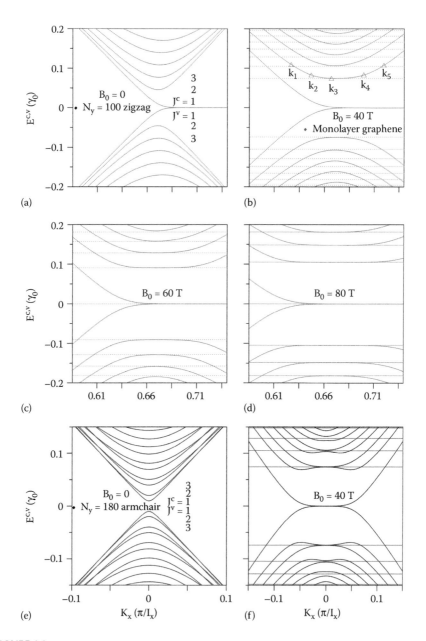

FIGURE 6.9
Band structures of an $N_y = 100$ zigzag nanoribbon at (a) $B_0 = 0$, (b) $B_0 = 40$ T, (c) $B_0 = 60$ T and (d) $B_0 = 80$ T, and those of an $N_y = 180$ armchair nanoribbon at (e) $B_0 = 0$ and (f) $B_0 = 40$ T. The red lines represent the LL energies of monolayer graphene. $J^{c,v}$ is the subband index, as measured from the Fermi level.

dispersionless relations simultaneously, for example, $J^{c,v}=2$ and 3. A sufficiently large nanoribbon width, being comparable to or longer than the magnetic length (l_B), accounts for the creation of QLLs. With a further increase of field strength, the dispersionless k_x-ranges become wider, and the higher-energy subbands might be transformed into QLLs, such as the magneto-electronic structures in Figure 6.6c and d at $B_0=60$ and 80 T, respectively. Specifically, the initial energies of QLLs will approach those of Landau levels LLs in monolayer graphene (red lines; the second term in Equation 3.2). These clearly indicate that electronic states are gradually quantized into Landau modes from lower-energy subbands with the increasing B_0 as they have smaller kinetic energies. It is relatively easy to observe dispersionless Landau states under enhanced field strength and the extension of nanoribbon width. The formation centers of QLLs, respectively, correspond to $k_x=2/3$ and 0 for zigzag and armchair nanoribbons (Figure 6.9b and f at 40 T). It should be noted that two neighboring subbands of the latter will gradually approach each other with the increase of B_0, covering the gradual couplings of ($J^c=1, J^v=1$), ($J^c=1, J^v=1$) and (J^v, J^v+1). As a result, the magneto-electronic structures of armchair nanoribbons might display splitting of the QLLs. There exist the extra band-edge states and energy gap is vanishing under a strong magnetic field.

The DOSs of 1D energy subbands present a lot of asymmetric peaks in the square-root form, as clearly indicated in Figure 6.10. The peak height, which is inversely proportional to the square root of subband curvature, grows as state energy increases. It is remarkable that the peak spacings are almost uniform in a zigzag system (Figure 6.10a), but non-uniform in an armchair one (Figure 6.10e). Specifically, the former has a pair of merged peaks near $E = 0$ or an obvious symmetric peak there (inset in Figure 6.10a). The low-lying asymmetric peaks become delta-function-like symmetric peaks when the QLLs can be created by a B_0-field (Figure 6.10b through d and f). Furthermore, their heights decline quickly with the increase of QLL energy, reflecting the diminishing of the dispersionless k_x-ranges (Figure 6.9). Under the increasing field strength, all the peak energies are enhanced for zigzag systems, while a simple dependence is absent for armchair ones [172,173]. As a result, the QLLs of the former will recover to become LLs of monolayer graphene if the magnetic field plays a dominating role (Figure 6.9b through d). The latter could exhibit the complex structures covering single and double peaks.

6.4 Magneto-Optical Spectra of Graphene Nanoribbons

When electrons are confined in finite-width nanoribbons, their spatial distributions reveal them as regular standing waves (Figure 6.11). The oscillatory patterns, with the specific number of nodes, are very sensitive to edge

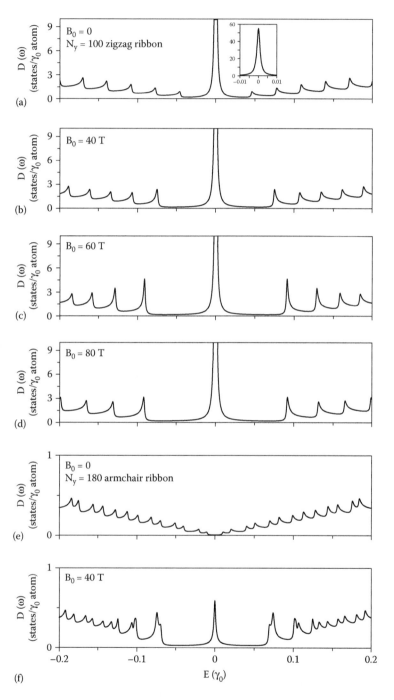

FIGURE 6.10
Density of states corresponding to electronic structures in Figure 6.9.

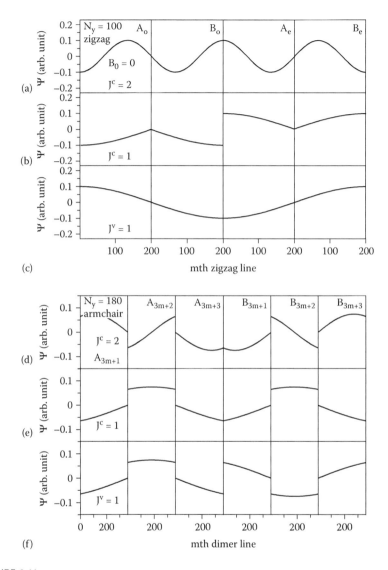

FIGURE 6.11
Subenvelope functions of an $N_y = 100$ zigzag nanoribbon for (a) $J^c = 2$, (b) $J^c = 1$ and (c) $J^v = 1$, and those of an $N_y = 180$ armchair system for (d) $J^c = 2$, (e) $J^c = 1$ and (f) $J^v = 1$.

structures, sublattices, state energies and wave vectors. They are quite different from those in carbon nanotubes (Figure 6.4), and so too are the optical properties. For zigzag nanoribbons, the wave functions can be decomposed into subenvelope functions on the A and B sublattices at the odd and even zigzag lines (Figure 6.11a through c). The spatial distributions of wave functions present the alternative change between symmetric and anti-symmetric forms as the subband index increases (more detailed relations in [41]). On the other hand, those of armchair systems lie at the 3m-, (3m + 1)- and (3m + 2)-th dimer lines of the A and B sublattices (Figure 6.11d through f). There are two kinds of unique relations. For a specific subband, the subenvelope functions on two sublattices possess the same phase or the phase difference of π. Furthermore, the valence and conduction subbands, with the identical index, exhibit a similar phase relation in the subenvelope functions of the A (*B*) sublattice. It is very significant that the special relations in wave functions are edge-dependent and thus dominate the distinct selection rules. However, the main features of the angular-momentum-dominated standing waves in carbon nanotubes (Figure 6.5) are independent of the chiral angle.

A dominating magnetic field can thoroughly alter the characteristics of wave functions, regardless of the edge structures. In general, the standing waves are changed into well-behaved LL wave functions, as clearly indicated in Figure 6.12a through e for a zigzag system. The $k_x = 2/3$ states present the symmetric or anti-symmetric distributions around the nanoribbon center. They are identical to those of monolayer graphene (Figure 3.4b) except that the first pair of QLLs possesses edge-localized distributions (Figure 6.12c and d). With the different wave vectors, the spatial distributions, revealed in Figure 6.12f through j, become asymmetric. Furthermore, the number of zero points might be lost as the wave vectors are far away from 2/3. Apparently, the competition between magnetic quantization and finite-size confinement will lead to coexistent behavior, in which LL modes and standing waves, respectively, correspond to low- and high-lying electronic states. This unusual feature could be observed by changing the B_0-field strength or nanoribbon width.

Graphene nanoribbons possess edge-dependent optical selection rules, as shown in Figure 6.13. The optical vertical excitations arise from the interband transitions of the J^v-th valence band and the J^c-th conduction band. For zigzag nanoribbons, a lot of asymmetric absorption peaks in the square-root form are characterized by the selection rule of $\Delta J = 2I + 1$ (Figure 6.13a). The strong absorption peaks might come from the multichannel excitations simultaneously, especially the higher-frequency ones. This rule could be directly derived from the non-vanishing velocity matrix elements (Equation 2.38) by using the special relations in the subenvelope functions (detailed calculations in [41]). On the other side, the available excitation channels in armchair systems agree with the $\Delta J = 0$ rule, so the number of absorption peaks is reduced. The valence and conduction bands, with the identical J, could also be revealed as the prominent absorption peaks of any carbon nanotubes (Figure 6.6), while the angular momentum is conserved during the vertical transitions.

The unusual transformation between the edge- and QLL-dominated absorption peaks could be presented in the variation of magnetic field strength (or nanoribbon width), as clearly shown in Figure 6.14a through c. The QLL wave functions have well-behaved spatial symmetry, so that the effective optical transitions associated with the symmetric absorption peaks are governed by the QLL-dependent selection rule. At a lower frequency,

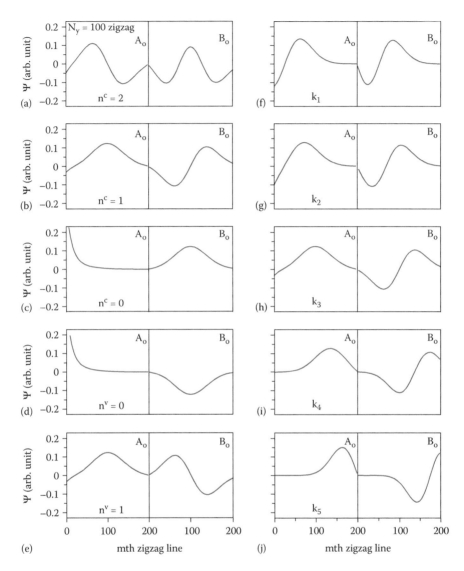

FIGURE 6.12

Magnetic subenvelope functions of an $N_y=100$ zigzag nanoribbon at $B_0=40$ T and $k_x=2/3$ for (a)–(c) $n^c=2$–0 and (e)–(f) $n^v=0$–1. Also shown are those of (f)–(j) $n^c=1$ at various k_x's (triangles in Figure 6.9b).

the magneto-absorption peaks of $\Delta J \neq \pm 1$ are absent or become very weak; furthermore, the $\Delta J = \pm 1$ rule is equivalent to the $\Delta n = \pm 1$ rule in monolayer graphene (red curve in Figure 6.14a). This is independent of edge structures; that is, the number, frequency and intensity of prominent absorption peaks are identical for zigzag and armchair nanoribbons with almost the same width (Figure 6.14a and d). Such peaks have a symmetric form and

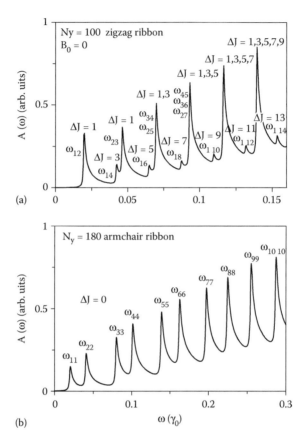

FIGURE 6.13

Optical absorption spectra of (a) $N_y = 100$ zigzag and (b) $N_y = 180$ armchair nanoribbons. Two subscripts in absorption peak frequency represent indexes of initial valence and final conduction bands.

stronger intensities at high field strengths (Figure 6.14c and d). In addition, some extra lower-absorption peaks are revealed by the $J^{c,v} = 1$ QLLs of the zigzag nanoribbons (e.g., blue circles in Figure 6.14a) because of the impure LL wavefunctions (Figure 6.12c and d), and the splitting QLLs of the armchair nanoribbons could create the double-peak structure (blue circles in Figure 6.14d). Concerning the higher-frequency asymmetric peaks (the right-hand side of the gray-dashed vertical line), there are more prominent structures, especially for complex absorption peaks in armchair nanoribbons (Figure 6.14d). They originate from the band-edge states of the parabolic valence and conduction bands, being characterized by the strong competition of the edge- and QLL-dependent selection rules.

    The complicated relations between lateral confinement, magnetic quantization and dimension deserve closer investigation. They could be understood from the initial six prominent magneto-absorption peaks, as clearly revealed

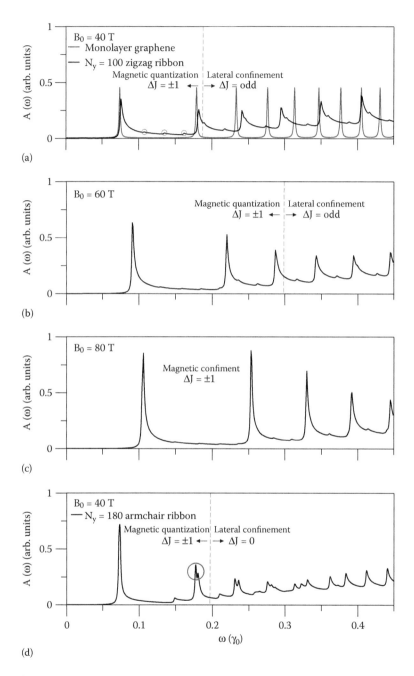

FIGURE 6.14

Magneto-optical absorption spectra for an $N_y=100$ zigzag nanoribbon at $B_0=$ (a) 40 T, (b) 60 T and (c) 80 T, and an $N_y=180$ armchair nanoribbon at (d) $B_0=40$ T. Also shown in (a) is the magneto-optical absorption spectra of monolayer graphene (red curve).

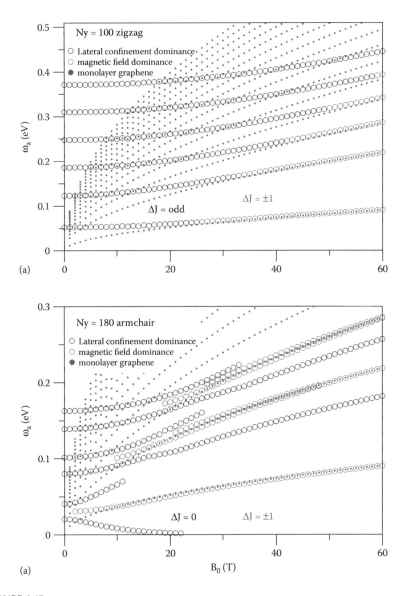

FIGURE 6.15
Magneto-absorption frequencies of the initial six prominent peaks for (a) $N_y = 100$ zigzag and (b) $N_y = 180$ armchair graphene nanoribbons. The red dots are the inter-LL optical excitation frequencies of monolayer graphene.

in Figure 6.15, for their B_0-dependent frequencies. Zigzag nanoribbons exhibit monotonic magnetic field dependence (open circles in Figure 6.15a). In general, peak frequencies grow with the increasing B_0 in the absence of a specific relation. They are very different from those of monolayer graphene (red dots) except for those with sufficiently high field strength. At a low B_0, the lateral confinement (black circles) dominates the frequency, intensity and form of absorption peaks, for example, the higher frequencies compared with 2D results. The magnetic quantization will also be seriously suppressed when the field strength is over a critical magnetic field (red circles). The number of prominent peaks remains the same during the variation of B_0, as the ΔJ = odd-induced peaks (Figure 6.13a) cover the QLL-created ones. However, absorption peaks in armchair systems might come to exist or disappear frequently as B_0 gradually increases (Figure 6.15b), being closely related to QLL splitting and the high competition between two selection rules. Obviously, a simple B_0-dependence of magneto-absorption frequencies is absent. The zero-field absorption peaks of $\Delta J = 0$ (black circles; Figure 6.13b) are not consistent with the $\Delta J = 0$ rule due to the QLLs, so their intensities decay rapidly and vanish after the critical B_0's. On the other hand, the new peaks initiated by strong magnetic quantization might appear in the double-peak structures (two close red-circled curves) or the merged ones. They may behave as inter-LL peaks in 2D monolayers (red dots) only at higher field strengths. Up to now, there have been no optical or magneto-optical measurements of graphene nanoribbons. However, the theoretical predictions of edge- and QLL-dominated selection rules could be verified by optical spectroscopies, as has been done for graphite, layered graphenes and carbon nanotubes.

6.5 Comparisons and Applications

The distinct stacking symmetries in three kinds of layered graphites have created diverse and novel physical phenomena. The critical differences cover electronic and optical properties in the absence/presence of magnetic quantization. The AA-, AB- and ABC-stacked graphites, respectively, have one, two and one pairs of π-electronic valence and conduction bands, in which the k_z-dependent bandwidths are ~1 eV, ~0.1 – 0.2 eV and ~0.01 eV, and the band-overlap widths near the Fermi level behave similarly. The carrier density of free electrons and holes is highest in simple hexagonal graphite, while it is lowest in rhombohedral graphite. The band structures of these three systems could be regarded as the 3D vertical Dirac cone, the composite of monolayer- and AB-bilayer-like cones and the spiral cone structure. The 3D energy dispersions determine the DOS characteristics, such as the semimetallic behavior and VHS-induced distinct structures. The band-dominated optical spectra exhibit dimension- and stacking-dependent characteristics at low and middle frequencies, including the frequency, number, form and spectral width of the special absorption structures. The main features of zero-field band structures are directly reflected in the magnetic Landau subbands (LSs). In general, the band-edge states of 1D parabolic LSs are shown as many asymmetric peaks in DOS. The AA system exhibits a lot of valence and conduction LSs intersecting with the Fermi level under specific energy spacings, while the ABC system only presents one crossing LS of $n^{c,v}=0$. Both of them possess monolayer-like subenvelope functions and $\sqrt{B_0}$-dependent energy spectra. However, the AB system has two groups of LSs with normal and perturbed modes, leading to the coexistence of crossing and anti-crossing behaviors. Only the initial two LSs of the first group cross the Fermi level. The B_0-dependence of LS energies is sensitive to the dimension-induced k_z, such as the square-root and linear dependences at $k_z=0$ and π (K and H points), respectively. Furthermore, the magnetic subenvelope functions might dramatically change during the variation of k_z. Specifically, even without any crossings and anti-crossings, the rich and unique magneto-optical spectra are revealed by well-behaved LSs in AA-stacked graphite, including the intraband and interband inter-LS vertical excitations, Fermi-momenta-induced absorption peaks, non-uniform peak intensity, multichannel threshold peak, intraband two-channel peaks and interband double-peak structures at distinct frequency ranges, discontinuous B_0-dependence of the initial interband channel, and the beating feature related to the vertical Dirac cones. Magneto-absorption peaks agree with the monolayer selection rule of $\Delta n=\pm 1$ for AA-, ABC- and AB-stacked graphites. The second stacking can create monolayer-like characteristics, such as interband transitions, non-composite symmetric peaks due to the same contribution of K and H points, almost uniform intensity and pure $\sqrt{B_0}$-dependence of absorption frequency. As for AB-stacked graphite, there are four categories of interband inter-LS excitations arising from the same or different groups. The strong asymmetric peaks might appear at identical frequency ranges and thus exhibit very complex absorption structures. Their main features are rather different for the K- and H-point vertical excitations, respectively, leading to bilayer- and monolayer-like absorption frequencies. Moreover, some extra peaks come to exist under the LS anti-crossings. The aforementioned differences could be examined using the experimental measurements of angle-resolved photoemission spectroscopy (ARPES), STS and optical spectroscopies on energy bands; DOS; and absorption spectra, respectively.

The dimensional crossover of the essential properties occurs in AA- and AB-stacked layered graphenes as the layer number gradually grows. The N-layer AA stacking has a vertical multi-Dirac cone structure that is distributed within the k_z-dependent bandwidth of simple hexagonal graphite. It is a semimetal under the overlap of valence and conduction bands. However, an optical gap is induced in an N-even system, as the absorption spectrum is only a combination of N intra-Dirac-cone vertical excitations. Under a perpendicular magnetic field, the magneto-optical gap comes from a forbidden transition region related to the intragroup LLs, and it is greatly enhanced by the increasing field strength. The magneto-threshold channels dramatically change with the increasing B_0, as their intensities are about half that of the others. The low-lying Dirac-cone structure, magnetically quantized states, special structures in DOS and absorption peaks are expected to approach those of AA-stacked graphite for $N > 30$. As for AB-stacked graphene, the band structure of an N-odd system resembles a hybridization of a massless Dirac cone and massive parabolic dispersions, while that of an N-even system consists of only pairs of parabolic subbands. When N is very large, the mono-layer- and bilayer-like states are expected to correspond to the H and K point in Bernal graphite, respectively. The excitation channels are only allowed between the respective monolayer-like subbands or bilayer-like subbands; the magneto-optical selection rule is also applicable to all inter-LL transi-tions. With the increase of layer number, the LL anti-crossings happen more frequently compared to those in Bernal graphite. Electron-hole asymmetry-induced twin-peak structures are revealed in both multilayer and bulk sys-tems. However, in a magneto-optical spectrum, the measured profiles of the B_0-dominated peaks, including threshold channel, intensity, spacing and frequency, could be used to distinguish the stacking layer, configuration and dimension. On the other hand, the dimensional crossover is rarely observed in ABC stacking, mainly owing to the distinct lattice symmetries in 3D and 2D systems. For bulk graphite, a primitive unit cell, rhombohedron, has a bipar-ticle lattice symmetry. The low-energy electronic properties are described by the 3D anisotropic spiral Dirac cones. Specifically, the ABC-stacked graphene possesses surface-localized and sombrero-shaped subbands, irrelevant to rhombohedral graphite. The zero-field absorption spectrum is contributed by the $N \times N$ excitation channels of the N pairs of energy bands, in which the low-frequency region is dominated by the surface-localized states. The magneto-optical spectrum consists of $N \times N$ groups of inter-LL absorption peaks, each of which displays characteristic B_0-dependence regarding the frequency, intensity and number. Furthermore, the frequent anti-crossing of LLs due to the sombrero-shaped subbands leads to extra peaks that have abnormal relations with field strength.

The significant effects due to lateral quantum confinement and magnetic field clearly illustrate the dimension-diversified essential properties. The periodical boundary condition induces the decoupled angular-momentum states in carbon nanotubes, but the open boundary condition cannot create a transverse quantum number in a graphene nanoribbon. The former pos-sesses sine/cosine standing waves along the azimuthal direction, regardless of radius and chirality. However, the unusual standing waves in the latter depend on edge structure, A and B sublattices and even zigzag and dimer lines. The distinct characteristics of subenvelope functions dominate the diverse selection rules, the conservation of angular momentum in carbon nanotubes, the same index of valence and conduction subbands in armchair nanoribbons and the index difference of odd integers in zigzag nanoribbons. All the special structures in DOS and absorption spectra are presented in the asymmetric peaks of the square-root form. The magnetic quantization in cylindrical nanotubes is mainly determined by the field direction and strength. A parallel magnetic field leads to the shift of angular momentum or the destruction of double degeneracy, and the periodical Aharonov–Bohm effect in energy dispersions, band gaps and absorption peaks. However, the discrete angular momenta are coupled with one another in the presence of a perpendicular magnetic field, and QLLs rarely survive in cylindrical systems except for in very high field strengths and large radii. Apparently, optical spectra are dramatically altered by the magnetic field, which can cause, for instance, more splitting peaks with the same rule and some extra peaks without the optical rule under parallel and perpendicular magnetic fields, respectively. On the other hand, QLLs could exist in graphene nanoribbons if the width is sufficient for localized oscillatory distribution. Their dispersionless k_x-ranges can transform the asymmetric peaks of the DOS and absorption spectrum into symmetric ones. Moreover, the latter and the former, which correspond to the QLL- and edge-dependent selection rules, respectively, appear at the lower and higher frequencies.

Graphite, as well as graphite intercalation compounds, has been extensively developed for various applications for a long time as a result of its unique properties. Pristine graphite is a suitable precursor in the production of other carbon-related materials. It is the most stable allotrope of carbon. The covalent bondings between C atoms in the same layer can generate graphite's high temperature stability and excellent electrical and thermal conductivity. Owing to its superior mechanical properties, graphite fibers are frequently utilized for reinforced composite materials [283–286]. Graphite intercalation compounds, with their significant chemical bondings between adatoms and carbons, exhibit wide applications in electronics [287,288], energy storage [289] and electrochemistry [290,291]. Its most common usage in industry is as a low-cost electrode [292,293]. The high free electron (hole) density is generated by the intercalation of metal atoms (molecules) [282]. The metallic compounds can be used as superconductors with high transition temperatures, for example, 11.5 K for C_6Ca and C_6Yb [12,299,300]. They have also been applied in the production of photoelectrocatalytic, electrochemical and biomedical sensors [294–298]. Graphite composites display high conductivity for microbial fuel cell applications in electrochemistry [301–303]. The field-effect transistor (FET) sensor based on graphite oxide nanoparticle composites presents a low-cost advantage and possesses high selectivity and excellent stability [304,305].

Graphene-related materials have displayed high potential for electronic, photonic and optoelectronic applications, such as touchscreen panel devices [307,308], light-emitting diodes (LEDs) [309–313], solar cells [314–317], photodetectors [318–325] and photo-modulators [326,327]. The direct application of graphene in FETs is suppressed owing to its zero-gap nature. However, bilayer graphene can open a sizable and tunable bandgap by applying a gate voltage, which is appropriate for making large-area graphene FETs with extremely thin, short, high-speed channels [328–331]. The high transparency and flexibility of graphene could be utilized to design thin, light and delicate devices [309,310]. Graphene-based optical modulators possess great advantages in terms of a low operation voltage, fast modulation speed, small footprint and large optical bandwidth, compared with semiconductor modulators [326,327]. By tuning the Fermi level or the free-carrier concentration of graphene sheets, the modulator is operated over a broad wavelength range. As a result of the strong and rich interband optical transitions, the graphene detectors are suitable for application within a very wide energy spectrum, covering ultraviolet, visible, infrared and terahertz frequency ranges [326,327]. When layered graphenes are further doped with adatoms or molecules, they might change into gap-modulated semiconductors or metals. Specifically, graphene oxides and hydrogenated graphenes possess adatom-modulated energy gaps in a wide range of $0 < E_g < 4.0$ eV, depending on the concentration and distribution of adatoms [332,333]. The tunable and controllable electronic properties make them potential candidates for applications such as FETs [334–336], supercapacitors [337,338], sensors [339,340], photovoltaic devices [341–343] and light-emitting devices [343,344]. The alkali- and Al-induced high free carrier density might have high potential in future technological applications, for example, high-capacity batteries [345,346] and energy storage [347,348].

Different carbon nanomaterials have been used for a wide range of applications in electronic devices, such as 1D carbon nanotubes [306,351–353] and graphene nanoribbons [6,306], 2D few-layer graphenes [306] and 3D graphites [12,287–289,292–300]. Up to now, FETs based on semiconducting nanotubes and nanoribbons have been widely developed, mainly owing to their advantage of high mobility under low scatterings. In carbon nanotubes, the intrinsic 1D electronic structures, with the decoupled states of angular momenta, dominate the 1D quantized electrical properties, including the radius-dependent resistance, capacitance and inductance, which are diameter-dependent and responsible for the nonmonotonic dependence of the electrical mobility [349]. The first FETs made of carbon nanotubes were reported in 1998 [351,352], which have superior electrical properties in conducting channels through gate-voltage modulation. The nanoscaled carbon nanotubes are responsible for low scatterings and allow the gate to control the potential of the channel in ultrathin FETs, while suppressing short-channel effects [352,353]. Moreover, semiconducting nanotubes have wide applications in optoelectronic devices, such as electroluminescent light emitters [354,355], supercapacitors [356] and photodetectors [322–324]. In carbon nanotube FETs, electrons and holes confined on the cylindrical surface are driven toward each other by applying an appropriate source-drain bias voltage. The recombination of the two types of the excited carriers (electrons and holes) emits electroluminescence, and the photon-emission process is extensively utilized in LEDs [357–359]. The application of photodetectors is based on the electric current generated by the resonant excitations. On the other side, metallic carbon nanotubes could serve as high-performance interconnects in integrated electronic devices [360,361].

Semiconducting graphene nanoribbons could be directly used in FETs [6,362,363], as their electronic states are confined in a narrow width and have obvious energy spacings (gaps) [365,366]. Edge structures and ribbon widths lead to different 1D electronic and optical properties, for example, the strong dependence of wave function on the edge or center position. With the decreasing ribbon width, the carrier mobility is degraded by the edge boundary, while the potential barrier is enhanced for the conducting channels. The on/off ratio is improved for narrow ribbons when the temperature is sufficiently lowered [6,363,364]. Another promising application of graphene nanoribbons is polymer composite and electrode material for batteries [367–369] and supercapacitors [374–376]. The synthesis of graphene nanoribbon composite has produced an effective component to improve the electrochemical stability and enhanced specific capacity of electrode materials.

7

Concluding Remarks

This book presents a systematic review of the essential properties of simple hexagonal, Bernal and rhombohedral graphites. The generalized tight-binding model and gradient approximation method are developed to explore the electronic and optical properties under magnetic quantization. Furthermore, the effective-mass approximation can provide qualitative pictures and semi-quantitative results. A thorough comparison is made between 3D graphite, 2D graphenes, 1D graphene nanoribbons and carbon nanotubes by covering the dependence on the layer number, stacking configuration, dimension, width/radius, edge/chirality and boundary condition. This is useful in understanding dimensional crossover behavior. The calculated results agree with those from other theoretical calculations and are validated by experimental measurements, while most of the predictions require further detailed examinations. The theoretical framework is useful in promoting future studies on other layered materials, for example, Si- [177,377], Ge- [378,379], Sn- [385], P- [181,381] and Bi-related 2D and 3D materials [386]. Specifically, the generalized tight-binding model is suitable for solving critical Hamiltonians with multi-orbital bondings, spin-orbital couplings, interlayer atomic interactions, and external electric and magnetic fields [174–177,179–181]. This model could be combined with the single- and many-particle theories to comprehend other physical properties, such as Coulomb excitations [59,371,373] and transport properties [372].

The intralayer and interlayer atomic interactions of $2p_z$ orbitals account for the diverse essential properties in layered graphites. AA-stacked graphite has the highest density of free electrons and holes (the largest band overlap), the widest energy dispersions along $\hat{k}_z$, vertical Dirac-cone structures and many Landau subbands (LSs) that cross the Fermi level. These directly reflect the highest stacking symmetry, or the strongest interlayer interactions. The optical spectrum presents a shoulder structure and a prominent plateau structure at low and middle frequencies, respectively. Due to the unusual magnetic quantization, one group of valence and conduction LSs can create a multichannel threshold peak, some intraband two-channel peaks and many interband double-peak structures with beating phenomena. The unique oscillational magneto-absorption spectra have never been predicted or identified in previous studies on any materials. On the other hand, layered graphenes exhibit more low-frequency shoulder structures, and optical gaps in *N*-even cases. The finite-layer confinement effect almost vanishes for $N > 30$; that is, AA-stacked graphenes and graphite possess the same optical spectra at that point. The magneto-absorption peaks have a symmetric structure and a uniform intensity. The threshold intraband peak is non-well-behaved in the B_0-dependence, and some initial interband peaks are absent. These are closely related to the quantized Landau levels (LLs) from the multi-Dirac cones. It is also noted that all absorption peaks due to LLs and LSs agree with the selection rule of $\Delta n = \pm, 1$, and they all possess $\sqrt{B_0}$-dependences except for the multichannel ones.

Bernal graphite is intriguing for studies of massless and massive Dirac quasiparticles. The research interest in 2D graphenes is based on the properties of bulk graphite, which are deduced to represent the coexistence of 2D monolayer and bilayer essential properties at different k_z wave vectors. Both Bernal graphene and graphite exhibit the optical response of Dirac fermions regardless of external fields. The band structure displays a massless Dirac-like behavior in the vicinity of the H point, where the in-plane dispersion is linear and doubly degenerate to reflect the two isolated graphene sheets in the primitive unit cell. On the other hand, the in-plane energy dispersions near the K point resemble massive Dirac-like behavior, which is specific to AB-stacked bilayer graphene. These monolayer-like and bilayer-like energy dispersions have indeed been observed by angle-resolved photoemission spectroscopy (ARPES) [71–75]. The field evolution of the 1D LSs is depicted in the Peierls tight-binding model. Under a strong magnetic field, the crossing of the low-lying LSs near the Fermi level gives rise to the change of semimetallic Bernal graphite into a zero-gap semiconductor. In the scale of B_0, two series of square-root divergent peaks with linear and square-root dependences are verified by scanning tunneling spectroscopy (STS), and these peaks account for the monolayer-like and bilayer-like Landau states that accumulate at the band edges of the LSs at the K and H points, respectively. Furthermore, depending on the curvatures of the LSs, the measured intensities of their own respective peaks are consistent with the theoretical calculations. The magneto-optical properties elucidated in the framework of the tight-binding model reveal far more significant results than the results derived from a simplified effective-mass approximation. This provides clarity to the information on the graphene-like properties in graphites and true epitaxial graphenes.

In the vicinity of the H and K points, the inter-LS channels are the dominant contributors to the magneto-absorption spectrum, the spectral intensity of which is determined by the density of states (DOS) intensity and the dipole transition probability. The monolayer-like and bilayer-like absorption spectra are predicted to coexist in the bulk spectrum following the characteristic magnetic field frequency dependences $\propto \sqrt{B_0}$ and $\propto B_0$. The measured Fermi velocity can be used to interpret γ_0 and γ_1 from the H-point and K-point channels; the deduced values in graphite match those in few-layer graphenes. Infrared magneto-absorption spectroscopies have confirmed the splitting of the absorption peaks for optical transition channels near the K and H points. However, the observability of the peak splittings in optical spectroscopy depends on the competition between the magnetic field, ambient temperature and experiment resolution. These main features are very useful in identifying the stacking configurations and dimensionality of systems from experimental measurements. The splitting at the K point is attributed to electron-hole asymmetry, reflecting the inherent complexity of the full interactions in the Slonczewski–Weiss–McClure (SWM) model. However, the splitting at the H point is still under debate as it does not result from Dirac-cone asymmetry but might originate from spin-orbital coupling, the anti-crossing of LSs or parallel magnetic flux. These results require a more elaborated model and better experimental verifications. In addition, the inconspicuous peaks, coming from the band-edge states of the anti-crossing LSs near the H point, could possibly be observed with the extra peak intensities enhanced by the degree of the hybridization of the LSs.

The ABC-stacking configuration has different point-group symmetries for the corresponding 2D and 3D structures, leading to distinct characteristics of electronic properties and optical spectra. For example, the massless Dirac quasiparticles are preserved in ABC-stacked graphite, and the bulk stack is topologically nontrivial for the existence of surface-localized states. The massless Dirac characteristics are even more obvious than those of AA- and AB-stacked graphites, because the energy dispersion dependence on k_z is weaker than that in Bernal graphite and simple hexagonal graphite by one or two orders of magnitude. The low-energy electronic and optical properties are reviewed in both rhombohedral and hexagonal unit cells, which are, respectively, built from two and eight sublattices. The former with $R\bar{3}m$ symmetry is the primitive unit cell of ABC-stacked graphite, whereas the latter with $P3$ symmetry is chosen to represent AA- and AB-stacked graphite for the sake of convenience. They provide the same physical results, but due to the zone-folding effect in the latter, the former is more appropriate to comprehend the evolution of the Dirac cone and magnetic quantization under the influence of different β_i's. In the minimal model with β_0 and β_1, the Dirac points rotate in a circular path with a constant radius of $\beta_1(v_0\hbar)^{-1}$ at the Fermi level. Each Dirac cone behaves as in a monolayer graphene with a Fermi velocity of $3\beta_0 b/2$, giving rise to a linearly increased intensity of ω in the absorption spectrum. Under a magnetic field, the corresponding LSs are completely reduced to 0D dispersionless LLs that are classified to one group, just like monolayer graphene. Such LLs induce 2D delta-function-like peaks in the DOS that are equal in intensity and that follow a simple square-root energy relationship $E^{c,v}(n^{c,v}) \propto \sqrt{n^{c,v}B_0}$. Also, the magneto-absorption spectrum is identical to that of monolayer graphene in which the 2D spectral peaks are in the sequence, $E(n^c \rightarrow n^v) \propto \sqrt{B_0}(\sqrt{n^c} + \sqrt{n^v})$, where $n^c - n^v = \pm 1$. These 2D characteristics based on the minimal model imply that the stacking effect is not demonstrated in the minimal model because of the lack of consideration for all interlayer atomic interactions.

The effect of β_i's beyond the minimal model are discussed in the generalized tight-bonding model. The Dirac cone becomes tilted and anisotropic, and, furthermore, it spirals around the corners of the first Brillouin zone (1st BZ) with a varying radius. These behaviors are believed to be caused by the influences of β_3 and β_4. The distortion of the isoenergy surfaces causes a deviation of linear absorption intensity in the low-energy region. With the knowledge of anisotropic energy dispersions, the magneto-electronic and magneto-optical properties are reviewed within semi-classical Onsager quantization and generalized Peierls tight-binding schemes. Both schemes are consistent in the low-energy region. The 1D k_z-dependent LSs are symmetric about the $n^{c,v} = 0$ LSs in the first BZ; however, the electron-hole symmetry is broken down because the dispersion of the $n^{c,v} = 0$ LSs moves according to the Dirac-point spiral. Based on the magneto selection rule $\Delta n = \pm 1$, the vertical transitions of a single channel are found to have approximately the same energy along K–H. Moreover, the energy deviation from the monolayer energy dependence directly indicates the distortion of Dirac cones. As a case for the experimental verification of the theoretically predicted optical and electronic properties of AB-stacked graphite, the spiral Dirac-cone structure can be verified by using the same experimental techniques, such as ARPES, STS and magneto-optical spectroscopy.

The reduced dimension in the transverse y-direction can greatly diversify the electronic properties and optical spectra of carbon-related systems, especially those of carbon nanotubes and graphene nanoribbons. The essential properties are very sensitive to the radius/width, chirality/edge and periodical/open boundary condition. As a result of the cylindrical symmetry, each carbon nanotube has angular-momentum-dependent electronic states, which are revealed as sine/cosine-form standing waves. The band-edge state energies and 1D energy dispersions strongly depend on radius and chirality, and so to do the frequency, number and intensity of 1D asymmetric absorption peaks. The $\Delta J = 0$ selection rule, which comes from the specific standing waves, represents the conservation of angular momentum during vertical excitations. This is independent of geometric structure. A parallel magnetic field induces the splitting of double degeneracy, the metal-semiconductor transition and the periodical Aharonov–Bohm effect, while a perpendicular one creates the coupling of distinct angular-momentum components and thus the destruction of the selection rule or extra absorption peaks. However, it is very difficult to observe the quasi-Landau level (QLL)-dominated essential properties, except for in cases with very high perpendicular magnetic fields or large carbon nanotubes.

As a result of the open boundary condition, graphene nanoribbons differ from carbon nanotubes due to the absence of a transverse quantum number, the edge-dominated standing waves, the edge-dependent selection rules, and the coexistence with the QLL-induced selection rule. The edge structure plays an important role in the existence of edge-localized states, uniform or non-uniform energy spacings and state degeneracy. Specifically, the subenvelope functions strongly rely on A and B sublattices, zigzag/dimer lines, state energies and wave vectors. For zigzag and armchair nanoribbons, one and two special relations are, respectively, presented in valence and conduction subbands, which are responsible for the $\Delta J = 2I+1$ and 0 selection rules of the zero-field optical spectra. The magnetic QLLs, being similar to LLs in monolayer graphene, are mainly determined by the competition between the width and magneto-length. They can exhibit lower-frequency symmetric absorption peaks with the $\Delta J = \pm 1$ selection rule. Furthermore, the asymmetric absorption peaks associated with the edge-dependent selection rule (the specific parabolic subbands) can survive at higher frequencies. The transformation between these two types of absorption peaks is clearly revealed in B_0-dependent magneto-optical spectra. It is relatively easily observed in zigzag nanoribbons compared with armchair systems. The latter can present very complicated peak structures because of the strong competition between $\Delta J = 0$ and ± 1 rules.

Part of the theoretical calculations agree with the experimental measurements. ARPES has identified the 3D energy bands of Bernal graphite [71–75], Dirac cone structure in monolayer graphene [233], two/three linear valence bands in bilayer/trilayer AA stacking [50,51], two pairs of parabolic bands in bilayer AB stacking [78,79], linear and parabolic bands in trilayer ABA stacking [78,80], partially flat, sombrero-shaped and linear bands in trilayer ABC stacking [80] and 1D parabolic subbands and energy gaps in graphene nanoribbons [163,232]. Similar ARPES examinations could be done for the electronic structures of simple hexagonal and rhombohedral graphites, and carbon nanotubes. The STS confirmations of the DOS characteristics cover a finite value at the Fermi level (the semimetallic behavior) and the bilayer- and monolayer-like LS energy spectra in Bernal graphite [71–75]; the V-shaped structure vanishing at $E=0$ and the $\sqrt{B_0}$-dependent LL energies in monolayer graphene; a special structure at $E \sim 0.3$ eV and the linear B_0-dependence (linear and square-root dependences) for LL energies in bilayer (trilayer) AB stacking [130,235–238]; a prominent peak near E_F arising from flat bands in trilayer ABC stacking [128–130]; the radius- and chirality-enriched energy gaps and asymmetric prominent peaks in carbon nanotubes [142,207,208]; and the confinement-induced band gaps in graphene nanoribbons [164–167]. The other prominent magneto-electronic structures in DOS, which are presented by AA- and ABC-stacked graphene/graphite, carbon nanotubes and graphene nanoribbons, deserve closer experimental verifications. As to the geometry-diversified electronic excitations, the optical/magneto-optical measurements have confirmed the π-electronic strong peak at middle frequency and the K- and H-dominated magneto-absorption peaks of the inter-LS transitions in AB-stacked graphite [85,89–93]; the low-frequency shoulder structure and the π peaks and the monolayer- and bilayer-like inter-LL absorption frequencies in few-layer AB stackings [20,86,87,131,255,256]; the two low-frequency characteristic peaks in trilayer ABC stacking [131,272]; and the radius-, chirality- and magnetic-field-dependent absorption peaks in carbon nanotubes [149]. Furthermore, converted absorption frequencies for the lowest sombrero-shaped band have been observed by magneto-Raman spectroscopy in ABC-stacked graphene of up to 15 layers [139]. Optical spectroscopies could be further used to check the intra- and inter-LS absorption peaks and the beating spectra in AA-stacked graphite, the transitions of intra-Dirac cones and intra-LL groups in few-layer AA stackings, the monotonic/complex magneto-excitation spectra in ABC-stacked graphite/graphenes and the edge- and QLL-dependent selection rules in graphene nanoribbons.

The various geometric structures and diverse intrinsic properties clearly indicate that graphite-related systems are suitable for the development of basic and applied sciences. Furthermore, the chemical doping of atoms and molecules could greatly enhance the application ranges. Pristine graphite is a suitable precursor for other carbon-related materials. Graphite fibers are frequently used as reinforced materials because of their excellent mechanical properties [283–286]. Graphite intercalation compounds, with tunable carrier densities, could serve as electrodes [287,288]; superconductors [12,299,300]; photoelectrocatalytic, electrochemical and biomedical sensors [294–298]; and microbial fuel cells [301–303]. Layered graphenes and their compounds, which possess rich and controllable electronic and optical properties, are expected to present a wide range of applications, such as field-effect transistors (FETs) [328–331], photodetectors [326,327], optical modulators [326,327], solar cells [314–317], touchscreen panel devices [307,308], various sensors [339,340], high-capacity batteries [345,346], supercapacitors [337,338] and energy storage [347,348]. The dimension-enriched essential properties in 1D carbon nanotubes have high potential in FETs [351–353], light-emitting diodes (LEDs) [357–359], electroluminescent light emitters [354,355], photodetectors [322–324] and high-performance interconnects [360,361]. Moreover, the finite-size effects of graphene nanoribbons can be used to develop nanoscale devices, for example, FETs [6,362–364].

The current work is closely related to the layered materials, with various lattice symmetries, planar/curved structures, stacking configurations, layer numbers and dimensions. The emergent group-IV 2D materials, which cover graphene, silicene [177,377], germanene [378,379], tinene [380] and monolayer Pb [381], are high-potential candidates for studying rich and unique physical, chemical and material phenomena. Such systems possess a lot of intrinsic properties in terms of lattice symmetries, planar or buckled structures, intralayer and interlayer atomic interactions, single- or multi-orbital chemical bondings [181], distinct site energies and spin-orbital couplings. The complicated relations between the significantly important interactions are expected to create the critical Hamiltonian and thus greatly diversify the essential properties. The generalized tight-binding model, which is reliable under uniform/non-uniform magnetic and electric fields [174–177,179–181], deserves further developments to make thorough and systematic investigations, especially for the diversities among five-layer systems. All the atomic interactions and external fields could be included in the calculations on electronic structures and optical properties simultaneously. The diverse phenomena in group-IV layered systems might become mainstream research topics in the near future, such as orbital-, spin- and valley-dominated magnetic quantizations; optical and magneto-optical selection rules; dimensional crossovers; adatom/molecule doping-induced energy gaps or free-carrier densities; stacking-modulated Dirac-cone structures and quantum Hall conductivities; and element-dependent plasmon modes and Landau dampings.

The combination of the generalized tight-binding model with the static Kubo formula is suitable for studying the quantum Hall effect (QHE) in layered materials. It could provide reliable LL energy spectra and wave functions, even under complicated anti-crossing behaviors. As a result, the available inter-LL transitions for the QHE and the selection rules are obtained exactly. The study of the bilayer and trilayer graphenes shows that the various stacking configurations greatly diversify the quantum transport properties [372]. The diverse features cover the non-integer conductivities; integer conductivities with distinct heights; LL-splitting-induced reduction and the complexity of quantum conductivity; zero or finite conductivity at the neutral point; and the well-like, staircase, composite and abnormal plateau structures in magnetic field dependencies. Similar studies on other 2D systems are expected to present more quantum phenomena.

As for the electronic Coulomb excitations, the delicate random-phase approximation has been successfully developed for 2D graphene systems, according to the layer-dependent subenvelope functions [59,371]. This point of view is the same as that of the generalized tight-binding model under various external fields. Thus, their combination could include intralayer and interlayer atomic interactions, intralayer and interlayer Coulomb interactions, and magnetic and electric fields simultaneously [59,61,177–180,371,373,382–384]. Up to now, systematic studies on single-particle and collective excitations (electron-hole pairs and plasmon modes) have been made for bilayer AA and AB stackings with/without magnetic quantization [59,382], monolayer graphene under a magnetic field [373], few-layer graphenes in the presence of an electric field [61,383,384], silicene with/without gate voltage or magnetic quantization [177–179] and germanene [180]. Such systems might exhibit unusual excitation phase diagrams associated with transferred momenta and energies, as they are never revealed in 2D electron gas systems. The many-particle phenomena arising from electron–electron interactions in emergent 2D materials are worthy of thorough investigations.

Problems

The following problems focus on the electronic and optical properties of graphene-related systems, and they can be solved by using the tight-binding model and the gradient approximation method in Chapter 2.

1. The essential properties of monolayer graphene are determined by the nearest-neighbor intralayer hopping integral of $2p_z$ orbitals. Calculate (1) energy bands and wave functions in the first Brillouin zone; and (2) velocity matrix elements during vertical optical excitations. Also, (3) discuss the main features of the velocity matrix elements near the Dirac point; and (4) examine whether the low-energy velocity matrix elements have a constant value.

2. Consider an AA-stacked bilayer graphene with the highest configuration symmetry. When the nearest-neighboring intralayer and the vertical interlayer hopping integrals are taken into account, show that (1) two pairs of Dirac cones can be characterized by analytic forms; (2) only the intra-Dirac-cone vertical transitions are available in absorption spectra by examining the linear superposition of the four tight-binding functions. (3) Show that the main features of Dirac cones in (1) and (2) are not affected by the gate voltage (a uniform perpendicular electric field).

3. Based on the low-energy expansions near the K point, (1) obtain the approximate Hamiltonian matrix elements for monolayer and AA bilayer stacking. Use creation and destruction operators to determine (2) the low-lying Landau levels (LLs) and well-behaved spatial distributions; (3) the specific magneto-optical selection rule; and (4) the threshold absorption frequency.

4. Monolayer graphene is present in a non-uniform perpendicular magnetic field. This field is assumed to be spatially modulated along the armchair direction in the cosine form, leading to an enlarged unit cell with $2N_B$ carbon atoms. Calculate (1) the vector potential; (2) the near-neighbor Peierls phases; and (3) the independent hopping integrals or Hamiltonian matrix elements. Similar calculations should be generalized to a uniform magnetic field accompanied by a modulated (4) electric or (5) magnetic field.

5. (1) Discuss the similarities and differences between AA-stacked graphenes and simple graphite for electronic and optical properties in the absence/presence of a uniform perpendicular magnetic field. (2) Apply the same investigations for AB-stacked graphenes and Bernal graphite.

6. For (1) trilayer ABA-stacked graphenes and (2) ABC-stacked graphenes, calculate the Hamiltonian matrix elements under a uniform magnetic field using the intralayer and interlayer hopping integrals in Chapter 2.

7. Explain why it is very difficult to observe the dimensional crossover behavior associated with ABC-stacked graphenes and rhombohedral graphite.

8. Carbon nanotubes possess the periodical boundary condition and the curvature effect. Evaluate the analytic π-electronic energy dispersions for (1) (P,P) armchair nanotubes; and (2) (P,0) zigzag nanotubes by using the nearest-neighbor hopping integrals provided in Section 6.1. Show that (3) the magneto-electronic band structures exhibit the Aharonov–Bohm effect in the presence of a parallel magnetic field. Also, discuss (4) the relation between energy gap and magnetic flux (ϕ) for armchair and zigzag systems by determining the lowest unoccupied state and the highest occupied state; and discuss (5) the magneto-optical selection rule under a parallel electric polarization.

9. An $N_y = 2$ zigzag nanoribbon is the smallest 1D system. Investigate (1) the energy bands and (2) wave functions of the Hamiltonian matrix elements with nearest-neighboring hopping integrals of $2p_z$ orbitals; and (3) the dependences on the magnetic flux through a hexagon.

10. Compare (1) the important differences between armchair (zigzag) nanotubes and zigzag (armchair) nanoribbons in terms of energy band, wave function, magnetic quantization and selection rule; and (2) propose concise physical pictures to explain them.

11. Discuss the evolution of magnetic quantization with dimension for AA, AB and ABC stackings by illustrating the sp²-bonding graphene-related systems.

References

1. R. Robertson, J. J. Fox and A. E. Martin, Two types of diamond, *Proc. R. Soc.* 1934, 232, 463–485.
2. J. D. Bernal, The structure of graphite, *Proc. R. Soc. London, Ser. A* 1924, 106, 749.
3. H. Lipson and A. R. Stokes, The structure of graphite, *Proc. R. Soc. London, Ser. A* 1942, 181, 101.
4. M. S. Dresselhaus and G. Dresselhaus, Intercalation compounds of graphite, *Adv. Phys.* 1980, 30, 139–326.
5. K. S. Novoselov, A. K. Geim, S. V. Morozov, D. Jiang, Y. Zhang, S. V. Dubonos, Griegorieva I. V. and Firsov A. A., Electric field effect in atomically thin carbon film, *Science* 2004, 306, 666–9.
6. X. Li, X. Wang, L. Zhang, S. Lee and H. Dai, Chemically derived, ultrasmooth graphene nanoribbon semiconductors, *Science* 2008, 319, 1229–32.
7. S. Iijima, Helical microtubules of graphitic carbon, *Nature* 1991, 354, 56–8.
8. J. Liu, A. G. Rinzler, H. J. Dai, J. H. Hafner, R. K. Bradley, P. J. Boul, A. Lu, T. Iverson, K. Shelimov, C. B. Huffman, F. Rodriguez-Macias, Y. S. Shon, T. R. Lee, D. T. Colbert and R. E. Smalley, Fullerene pipes, *Science* 1998, 280, 1253–6.
9. Z. Q. Li, E. A. Henriksen, Z. Jiang, Z. Hao, M. C. Martin, P. Kim, H. L. Stormer and D. N. Basov, Band structure asymmetry of bilayer graphene revealed by infrared spectroscopy, *Phys. Rev. Lett.* 2009, 102, 037403.
10. J. C. Blancon, M. Paillet, H. N. Tran, X. T. Than, S. A. Guebrou, A. Ayari, A. S. Miguel, N. M. Phan, A. A. Zahab and J. L. Sauvajol, Direct measurement of the absolute absorption spectrum of individual semiconducting single-wall carbon nanotubes, *Nat. Comm.* 2013, 4, 2542.
11. D. M. Otpmebs and H. F. Rase, Potassium graphites prepared by mixed-reaction technique, *Carbon* 1966, 4, 125–7.
12. N. Emery, C. Herold, J. F. Mareche and P. Lagrange, Synthesis and superconducting properties of CaC6, *Sci. Tech. Adv. Mater.* 2008, 9, 044102.
13. M. S. Dresselhaus and G. Dresselhaus, Intercalation compounds of graphite, *Adv. Phys.* 2002, 51, 1–186.
14. I. Belash, A. Bronnikov, O. Zharikov and A. Pal'nichenko, Superconductivity of graphite intercalation compound with lithium C^{2}Li, *Solid State Commun.* 1989, 69, 921.
15. C. Lee, X. Wei, J. W. Kysar and J. Hone, Measurement of the elastic properties and intrinsic strength of monolayer graphene, *Science* 2008, 321. 385–8.
16. A. H. Castro Neto, F. Guinea, N. M. R. Peres, K. S. Novoselov and A. K. Geim, The electronic properties of graphene, *Rev. Mod. Phys.* 2009, 81. 109–62.
17. E. McCann and M. Koshino, The electronic properties of bilayer graphene, *Rep. Prog. Phys.* 2013, 76, 056503.
18. Y. H. Ho, Y. H. Chiu, D. H. Lin, C. P. Chang and M. F. Lin, Magneto-optical selection rules in bilayer Bernal graphene, *ACS Nano* 2010, 4, 1465–72.
19. D. S. L. Abergel and V. I. Fal'ko, Optical and magneto-optical far-infrared properties of bilayer graphene, *Phys. Rev. B* 2007, 75, 155430.

20. M. Orlita, C. Faugeras, J. Borysiuk, J. M. Baranowski, W. Strupiński, M. Sprinkle, C. Berger, W. A. de Heer, D. M. Basko, G. Martinez and M. Potemski, Magneto-optics of bilayer inclusions in multilayered epitaxial graphene on the carbon face of SiC, *Phys. Rev. B* 2011, 83, 125302.

21. Z. Jiang, E. A. Henriksen, L. C. Tung, Y. J. Wang, M. E. Schwartz, M. Y. Han, P. Kim and H. L. Stormer., Infrared spectroscopy of Landau levels of graphene, *Phys. Rev. Lett.* 2007, 98, 197403.

22. P. Kuhne, V. Darakchieva, R. Yakimova, J. D. Tedesco, R. L. Myers-Ward, C. R. Eddy, Jr., D. K. Gaskill, C. M. Herzinger, J. A. Woollam, M. Schubert and T. Hofmann, Polarization selection rules for inter-Landau-level transitions in epitaxial graphene revealed by the infrared optical Hall effect, *Phys. Rev. Lett.* 2013, 111, 077402.

23. M. Orlita, C. Faugeras, P. Plochocka, P. Neugebauer, G. Martinez, D. K. Maude, A.-L. Barra, M. Sprinkle, C. Berger, W. A. de Heer and M. Potemski, Approaching the Dirac point in high-mobility multilayer epitaxial graphene, *Phys. Rev. Lett.* 2008, 101, 267601.

24. P. Plochocka, C. Faugeras, M. Orlita, M. L. Sadowski, G. Martinez, M. Potemski, M. O. Goerbig, J.-N. Fuchs, C. Berger and W. A. de Heer, High-energy limit of massless Dirac fermions in multilayer graphene using magneto-optical transmission spectroscopy, *Phys. Rev. Lett.* 2008, 100, 087401.

25. T. N. Do, P. H. Shih, C. P. Chang, C. Y. Lin and M. F. Lin, Rich magneto-absorption spectra of AAB-stacked trilayer graphene, *Phys. Chem. Chem. Phys.* 2016, 18, 17597.

26. C. Y. Lin, J. Y. Wu, Y. J. Ou, Y. H. Chiu and M. F. Lin, Magneto-electronic properties of multilayer graphenes, *Phys. Chem. Chem. Phys.* 2015, 17, 26008–35.

27. Y. H. Lai, J. H. Ho, C. P. Chang and M. F. Lin, Magnetoelectronic properties of bilayer Bernal graphene, *Phys. Rev. B* 2008, 77, 085426.

28. E. McCann and V. I. Fal'ko, Landau-level degeneracy and quantum hall effect in a graphite bilayer, *Phys. Rev. Lett.* 2006, 96, 086805.

29. H. Min and A. H. MacDonald, Chiral decomposition in the electronic structure of graphene multilayers, *Phys. Rev. B* 2008, 77, 155416.

30. S. H. R. Sena, J. M. Pereira, F. M. Peeters and G. A. Faria, Landau levels in asymmetric graphene trilayers, *Phys. Rev. B* 2011, 84, 205448.

31. C. Y. Lin, J. Y. Wu, Y. H. Chiu, C. P. Chang and M. F. Lin, Stacking-dependent magnetoelectronic properties in multilayer graphene, *Phys. Rev. B* 2014, 90, 205434.

32. Y. P. Lin, J. Wang, J. M. Lu, C. Y. Lin and M. F. Lin, Energy spectra of ABC-stacked trilayer graphene in magnetic and electric fields, *RSC Adv.* 2014, 4, 56552–60.

33. N. T. T. Tran, S. Y. Lin, O. E. Glukhova and M. F. Lin, Configuration-induced rich electronic properties of bilayer graphene, *J. Phys. Chem. C* 2015, 119, 10623–30.

34. T. N. Do, C. Y. Lin, Y. P. Lin, P. H. Shih and M. F. Lin, Configuration-enriched magnetoelectronic spectra of AAB-stacked trilayer graphene, *Carbon* 2015, 94, 619–32.

35. Y. K. Huang, S. C. Chen, Y. H. Ho, C. Y. Lin and M. F. Lin, Feature-rich magnetic quantization in sliding bilayer graphenes. *Sci. Rep.* 2014, 4, 7509.

36. C. Y. Lin, J. Y. Wu, Y. J. Ou, Y. H. Chiu and M. F. Lin, Magneto-electronic properties of multilayer graphenes, *Phys. Chem. Chem. Phys.* 2015, 17, 26008.

37. K. S. Novoselov, A. K. Geim, S. V. Morozov, D. Jiang, M. I. Katsnelson, I. V. Grigorieva, S. V. Dubonos and A. A. Firsov, Two-dimensional gas of massless Dirac fermions in graphene, *Nature* 2005, 438, 197–200.

38. Y. B. Zhang, Y. W. Tan, H. L. Stormer and P. Kim, Experimental observation of the quantum Hall effect and Berry's phase in graphene, *Nature* 2005, 438, 201–4.

39. K. S. Novoselov, E. McCann, S. V. Morozov, V. I. Fal'ko, M. I. Katsnelson, U. Zeitler, D. Jiang, F. Schedin and A. K. Geim, Unconventional quantum Hall effect and Berry's phase of 2π in bilayer graphene, *Nat. Phys.* 2006, 2, 177–80.

40. T. Taychatanapat, K. Watanabe, T. Taniguchi and P. J. Herrero, Quantum Hall effect and Landau-level crossing of Dirac fermions in trilayer graphene, *Nat. Phys.* 2008, 7, 621–5.

41. L. Zhang, Y. Zhang, J. Camacho, M. Khodas and I. Zaliznyak, The experimental observation of quantum Hall effect of $l = 3$ chiral quasiparticles in trilayer graphene, *Nat. Phys.* 2011, 7, 953–7.

42. X. Xia, J. Wang, F. Zhang, Z. D. Hu, C. Liu, X. Yan and L. Yuan, Multi-mode plasmonically induced transparency in dual coupled graphene-integrated ring resonators, *Plasmonics* 2015, 10, 1409–15.

43. F. H. L. Koppens, D. E. Chang and F. J. G. de Abajo, Graphene plasmonics: A platform for strong light-matter interactions, *Nano Lett.* 2011, 11, 3370–7.

44. J. Christensen, A. Manjavacas, S. Thongrattanasiri, F. H. L. Koppens and F. J. G. de Abajo, Graphene plasmon waveguiding and hybridization in individual and paired nanoribbons, *ACS Nano* 2012, 6, 431–40.

45. H. Yan, T. Low, W. Zhu, Y. Wu, M. Freitag, X. Li, F. Guinea, P. Avouris and F. Xia, Damping pathways of mid-infrared plasmons in graphene nanostructures, *Nat. Photonics* 2013, 7, 394–9.

46. J. C. Charlier, X. Gonze and J. P. Michenaud, First principles study of the stacking effect on the electronic properties of graphite, *Carbon* 1994, 32, 289.

47. J. K. Lee, S. C. Lee, J. P. Ahn, S. C. Kim, J. I. B. Wilson and P. John, The growth of AA graphite on (111) graphite, *J. Chem. Phys.* 2008, 129, 234709.

48. S. Horiuchi, T. Gotou, M. Fujiwara, R. Sotoaka, M. Hirata, K. Kimoto, T. Asaka, T. Yokosawa, Y. Matsui and K. Watanabe, Carbon nanofilm with a new structure and property, *Jpn. J. Appl. Phys.* 2003, 47, 1073.

49. J. Borysiuk, J. Soltys and J. Piechota, Stacking sequence dependence of graphene layers on SiC (eqn_0001.eps): Experimental and theoretical investigation, *J. Appl. Phys.* 2011, 109, 093523.

50. K. S. Kim, A. L. Walter, L. Moreschini, T. Seyller, K. Horn, E. Rotenberg and A. Bostwick, Coexisting massive and massless Dirac fermions in symmetry-broken bilayer graphene, *Nat. Mater.* 2013, 12, 887–92.

51. C. Bao, W. Yao, E. Wang, C. Chen, J. Avila, M. C. Asensio and S. Zhou, Stacking-dependent electronic structure of trilayer graphene resolved by nanospot angle-resolved photoemission spectroscopy, *Nano Lett.* 2017, 8, 1564.

52. I. Lobato and B. Partoens, Multiple Dirac particles in AA-stacked graphite and multilayers of graphene, *Phys. Rev. B* 2011, 83, 165429.

53. J. C. Charlier, J.-P. Michenaud, X. Gonze and J.-P. Vigneron, Tight-binding model for the electronic properties of simple hexagonal graphite, *Phys. Rev. B* 1991, 44, 13237.

54. C. L. Lu, C. P. Chang and M. F. Lin, Magneto-electronic properties of the AA- and ABC-stacked graphites, *Eur. Phys. J. B* 60, 161–9.

55. R. B. Chen and Y. H. Chiu, Landau subband and Landau level properties of AA-stacked graphene superlattice, *J. Nanosci. Nanotechnol.* 2012, 12, 2557–66.

56. Y. H. Ho, J. Y. Wu, R. B. Chen, Y. H. Chiu and M. F. Lin, Optical transitions between Landau levels: AA-stacked bilayer graphene, *Appl. Phys. Lett.* 2010, 97, 101905.

57. R. B. Chen, Y. H. Chiu and M. F. Lin, A theoretical evaluation of the magneto-optical properties of AA-stacked graphite, *Carbon* 2012, 54, 248.

58. R. B. Chen, Y. H. Chiu and M. F. Lin, Beating oscillations of magneto-optical spectra in simple hexagonal graphite, *Comput. Phys. Commun.* 2014, 189, 60.

59. J. H. Ho, C. L. Lu, C. C. Hwang, C. P. Chang and M. F. Lin, Coulomb excitations in AA- and AB-stacked bilayer graphites, *Phys. Rev. B* 2006, 74, 085406.

60. J. Y. Wu, G. Gumbs and M. F. Lin, Combined effect of stacking and magnetic field on plasmon excitations in bilayer graphene, *Phys. Rev. B* 2014, 89, 165407.

61. Y. C. Chuang, J. Y. Wu and M. F. Lin, Electric-field-induced plasmon in AA-stacked bilayer graphene, *Ann. Phys.* 2013, 339, 298.

62. R. B. Chen, C. W. Chiu and M. F. Lin, Magnetoplasmons in simple hexagonal graphite, *RSC Adv.* 2015, 5, 53736–40.

63. C. W. Chiu, F. L. Shyu, M. F. Lin, G. Gumbs and O. Roslyak, Anisotropy of £k-plasmon dispersion relation of AA-stacked graphite, *J. Phys. Soc. Jpn.* 2012, 81, 104703.

64. Z. Klusek, Investigations of splitting of the π bands in graphite by scanning tunneling spectroscopy, *Appl. Surf. Sci.* 1999, 151, 251.

65. G. Li, A. Luican and E. Y. Andrei, Scanning tunneling spectroscopy of graphene on graphite, Phys. Rev. Lett. 2009, 102, 176804.

66. Y. F. Hsu and G. Y. Guo, Anomalous integer quantum Hall effect in AA-stacked bilayer graphene, *Phys. Rev. B* 2010, 82, 165404.

67. A. I. Cocemasov, D. L. Nika and A. A. Balandin, Phonons in twisted bilayer graphene, *Phys. Rev. B* 2013, 88, 035428.

68. Z. Sun, A.-R. O. Raji, Y. Zhu, C. Xiang, Z. Yan, C. Kittrell, E. L. G. Samuel and J. M. Tour, Large-area Bernal-stacked bi-, tri-, and tetralayer graphene, *ACS Nano* 2012, 6, 9790–6.

69. P. Lauffer, K. Emtsev, R. Graupner, T. Seyller, L. Ley, S. Reshanov and H. Weber, Atomic and electronic structure of few-layer graphene on SiC (0001) studied with scanning tunneling microscopy and spectroscopy, *Phys. Rev. B* 2008, 77, 155426.

70. D. Pierucci, T. Brumme, J. C. Girard, M. Calandra, M. G. Silly, F. Sirotti, A. Barbier, F. Mauri and A. Ouerghi, Atomic and electronic structure of trilayer graphene/SiC(0001): Evidence of strong dependence on stacking sequence and charge transfer, *Sci. Rep.* 2016, 6, 33487.

71. A. Gruneis, C. Attaccalite, T. Pichler, V. Zabolotnyy, H. Shiozawa, S. L. Molodtsov, D. Inosov, A. Koitzsch, M. Knupfer, J. Schiessling, R. Follath, R. Weber, P. Rudolf, L. Wirtz and A. Rubio, Electron–electron correlation in graphite: A combined angle-resolved photoemission and first-principles study, *Phys. Rev. Lett.* 2008, 100, 037601.

72. C. M. Cheng, C. J. Hsu, J. L. Peng, C. H. Chen, J. Y. Yuh and K. D. Tsuei, Tight-binding parameters of graphite determined with angle-resolved photoemission spectra, *Appl. Surf. Sci.* 2015, 354, 229–34.

73. R. Kundu, P. Mishra, B. R. Sekhar, M. Maniraj and S. R. Barman, Electronic structure of single crystal and highly oriented pyrolytic graphite from ARPES and KRIPES, *Phys. B* 2012, 407, 827–32.

74. C. S. Leem, C. Kim, S. R. Park, M. K. Kim, H. J. Choi and C. Kim, High-resolution angle-resolved photoemission studies of quasiparticle dynamics in graphite, *Phys. Rev. B* 2009, 79, 125438.

75. S. Y. Zhou, G.-H. Gweon, J. Graf, A. V. Fedorov, C. D. Spataru, R. D. Diehl, Y. Kopelevich, D.-H. Lee, Steven G. Louie and A. Lanzara, First direct observation of Dirac fermions in graphite, *Nat. Phys.* 2006, 2, 595–9.

76. T. Ohta, A. Bostwick, J. L. McChesney, T. Seyller, K. Horn and E. Rotenberg, Interlayer interaction and electronic screening in multilayer graphene investigated with angle-resolved photoemission spectroscopy, *Phys. Rev. Lett.* 2007, 98, 206802.

77. T. Ohta, A. Bostwick, T. Seyller, K. Horn and E. Rotenberg, Controlling the electronic structure of bilayer graphene, *Science* 2006, 313, 951–4.

78. T. Ohta, A. Bostwick, J. L. McChesney, T. Seyller, K. Horn and E. Rotenberg, Interlayer interaction and electronic screening in multilayer graphene investigated with angle-resolved photoemission spectroscopy, *Phys. Rev. Lett.* 2007, 98, 206802.

79. T. Ohta, A. Bostwick, T. Seyller, K. Horn and E. Rotenberg, Controlling the electronic structure of bilayer graphene, *Science* 2006, 313, 951.

80. C. Coletti, S. Forti, A. Principi, K. V. Emtsev, A. A. Zakharov, K. M. Daniels, B. K. Daas, M. V. S. Chandrashekhar, T. Ouisse, D. Chaussende, A. H. MacDonald, M. Polini and U. Starke, Revealing the electronic band structure of trilayer graphene on SiC: An angle-resolved photoemission study, *Phys. Rev. B* 2013, 88, 155439.

81. G. Li and E. Y. Andrei, Observation of Landau levels of Dirac fermions in graphite, *Nat. Phys.* 2007, 3, 623–7.

82. T. Matsui, H. Kambara, Y. Niimi, K. Tagami, M. Tsukada and H. Fukuyama, STS observations of Landau levels at graphite surfaces, *Phys. Rev. Lett.* 2005, 94, 226403.

83. D. L. Miller, K. D. Kubista, G. M. Rutter, M. Ruan, W. A. de Heer, P. N. First and J. A. Stroscio, Observing the quantization of zero mass carriers in graphene, *Science* 2009, 324, 924–7.

84. L. J. Yin, S. Y. Li, J. B. Qiao, J. C. Nie and L. He, Landau quantization in graphene monolayer, Bernal bilayer, and Bernal trilayer on graphite surface, *Phys. Rev. B* 2015, 91, 115405.

85. E. A. Taft and H. R. Philipp, Optical properties of graphite, *Phys. Rev.* 1965, 138, A197.

86. K. F. Mak, J. Shan and T. F. Heinz, Seeing many-body effects in single- and few-layer graphene: Observation of two-dimensional saddle-point excitons, *Phys. Rev. Lett.* 2011, 106, 046401.

87. V. G. Kravets, A. N. Grigorenko, R. R. Nair, P. Blake, S. Anissimova, K. S. Novoselov and A. K. Geim, Spectroscopic ellipsometry of graphene and an exciton-shifted van Hove peak in absorption, *Phys. Rev. B* 2010, 81, 155413.

88. Y. H. Ho, Y. H. Chiu, W. P. Su and M. F. Lin, Magneto-absorption spectra of Bernal graphite, *Appl. Phys. Lett.* 2011, 99, 011914.

89. K.-C. Chuang, A. M. R. Baker and R. J. Nicholas, Magnetoabsorption study of Landau levels in graphite, *Phys. Rev. B* 2009, 80, 161410(R).

90. M. Orlita, C. Faugeras, J. M. Schneider, G. Martinez, D. K. Maude and M. Potemski, Graphite from the viewpoint of Landau level spectroscopy: An effective graphene bilayer and monolayer, *Phys. Rev. Lett.* 2009, 102, 166401.

91. N. A. Goncharuk, L. Nádvorník, C. Faugeras, M. Orlita and L. Smrčka, Infrared magnetospectroscopy of graphite in tilted fields, *Phys. Rev. B* 2012, 86, 155409.

92. M. Orlita, C. Faugeras, A.-L. Barra, G. Martinez, M. Potemski, D. M. Basko, M. S. Zholudev, F. Teppe, W. Knap, V. I. Gavrilenko, N. N. Mikhailov, S. A. Dvoretskii, P. Neugebauer, C. Berger and W. A. de Heer, Infrared magneto-spectroscopy of two-dimensional and three-dimensional massless fermions: A comparison, *J. App. Phys.* 2015, 117, 112803.

93. M. Orlita, C. Faugeras, G. Martinez, D. K. Maude, M. L. Sadowski and M. Potemski, Dirac fermions at the H Point of graphite: Magnetotransmission studies, *Phys. Rev. Lett.* 2008, 100, 136403.

94. W. W. Toyt and M. S. Dresselhaus, Minority carriers in graphite and the H-point magnetoreflection spectra, *Phys. Rev. B* 1977, 15, 4077.

95. K. F. Mak, M. Y. Sfeir, Y. Wu, C. H. Lui, J. A. Misewich and T. F. Heinz, Measurement of the optical conductivity of graphene, *Phys. Rev. Lett.* 2008, 101, 196405.

96. Z. Q. Li, S.-W. Tsai, W. J. Padilla, S. V. Dordevic, K. S. Burch, Y. J. Wang and D. N. Basov, Infrared probe of the anomalous magnetotransport of highly oriented pyrolytic graphite in the extreme quantum limit, *Phys. Rev. B* 2006, 74, 195404.

97. P. R. Wallace, The band theory of graphite, *Phys. Rev.* 1947, 71, 622.

98. J. C. Slonczewski and P. R. Weiss, Band structure of graphite, *Phys. Rev.* 1958, 109, 272.

99. J. W. McClure, Theory of diamagnetism of graphite, *Phys. Rev.* 1960, 119, 606.

100. Y. H. Ho, J. Wang, Y. H. Chiu, M. F. Lin and W. P. Su, Characterization of Landau subbands in graphite: A tight-binding study, *Phys. Rev. B* 2011, 83, 121201.

101. K. Nakao, Landau level structure and magnetic breakthrough in graphite, *J. Phys. Soc. Jpn.* 1976, 40, 761–8.

102. M. Inoue, Landau levels and cyclotron resonance in graphite, *J. Phys. Soc. Jpn.* 1962, 17, 808–19.

103. Q. Y. Lin, T. Q. Li, Z. J. Liu, Y. Song, L. L. He, Z. J. Hu, Q. G. Guo and H. Q. Ye, High-resolution TEM observations of isolated rhombohedral crystallites in graphite blocks, *Carbon* 2012, 50, 2369.

104. Z. Zhou, W. G. Bouwman, H. Schut and C. Pappas, Interpretation of X-ray diffraction patterns of (nuclear) graphite, *Carbon* 2014, 69, 17.

105. N. S. Saenko, The x-ray diffraction study of three-dimensional disordered network of nanographites: Experiment and theory, *Phys. Procedia.* 2012, 23, 102.

106. Y. Hishiyama and M. Nakamura, X-ray diffraction in oriented carbon films with turbostratic structure, *Carbon* 1995, 33, 1399.

107. P. Xu, Y. R. Yang, S. D. Barber, J. K. Schoelz, D. Qi, M. L. Ackerman, L. Bellaiche and P. M. Thibado, New scanning tunneling microscopy technique enables systematic study of the unique electronic transition from graphite to graphene, *Carbon* 2012, 50, 4633.

108. C. D. Zeinalipour-Yazdi and D. P. Pullman, A new interpretation of the scanning tunneling microscope image of graphite, *Chem. Phys.* 2008, 348, 233.

109. H. Kempa, P. Esquinazi and Y. Kopelevich, Integer quantum Hall effect in graphite, *Solid State Commun.* 2006, 138, 118–22.

110. N. Luiggi and M. Gómez, Rhombohedral graphite: Comparative study of the electronic properties, *J. Mol. Struct: THEOCHEM* 2009, 897, 118–28.

111. C. H. Ho, C. P. Chang, W. P. Su and M. F. Lin, Precessing anisotropic Dirac cone and Landau subbands along a nodal spiral, *New J. Phys.* 2013, 15, 053032.

112. C. H. Ho, C. P. Chang and M. F. Lin, Landau subband wavefunctions and chirality manifestation in rhombohedral graphite, *Solid State Commun.* 2014, 197, 11.

113. B. A. Bernevig, T. L. Hughes, S. Raghu and D. P. Arovas, Theory of the three-dimensional quantum hall effect in graphite, *Phys. Rev. Lett.* 2007, 99, 146804.

114. J. Hass, W. A. De Heer and E. H. Conrad, The growth and morphology of epitaxial multilayer graphene, *J. Phys. Condens. Matter* 2008, 20, 323202.

115. J. Coraux, A. T. N'Diaye, C. Busse and T. Michely, Structural coherency of graphene on Ir (111), *Nano Lett.* 2008, 8, 565.

116. A. Ismach, C. Druzgalski, S. Penwell, A. Schwartzberg, M. Zheng, A. Javey, J. Bokor and Y. Zhang, Direct chemical vapor deposition of graphene on dielectric surfaces, *Nano Lett.* 2010, 10, 1542.

117. H. J. Park, J. Meyer, S. Roth and V. Skkalov, Growth and properties of few-layer graphene prepared by chemical vapor deposition, *Carbon* 2010, 48, 1088.

118. A. Reina, X. Jia, J. Ho, D. Nezich, H. Son, V. Bulovic, M. S. Dresselhaus and J. Kong, Large area, few-layer graphene films on arbitrary substrates by chemical vapor deposition, *Nano Lett.* 2008, 9, 30.

119. S. J. Chae, F. Gune, K. K. Kim, E. S. Kim, G. H. Han, S. M. Kim, H.-J. Shin, S.-M. Yoon, J.-Y. Choi, M. H. Park, C. W. Yang, D. Pribat and Y. H. Lee, Synthesis of large-area graphene layers on poly-nickel substrate by chemical vapor deposition: Wrinkle formation, *Adv. Mater.* 2009, 21, 2328.

120. G. Zhao, J. Li, X. Ren, C. Chen and X. Wang, Few-layered graphene oxide nanosheets as superior sorbents for heavy metal ion pollution management. *Environ. Sci. Technol.* 2011, 45, 10454.

121. S. Stankovich, D. A. Dikin, R. D. Piner, K. A. Kohlhaas, A. Kleinhammes, Y. Jia, et al., Synthesis of graphene-based nanosheets via chemical reduction of exfoliated graphite oxide, *Carbon* 2007, 45, 1558.

122. C. Rao, K. Subrahmanyam, H. R. Matte, B. Abdulhakeem, A. Govindaraj, B. Das, P. Kumar, A. Ghosh and D. Late, A study of the synthetic methods and properties of graphenes, *Sci. Technol. Adv. Mater.* 2010, 11, 054502.

123. B. Qin, T. Zhang, H. Chen and Y. Ma, The growth mechanism of few-layer graphene in the arc discharge process, *Carbon* 2016, 102, 494.

124. Z. S. Wu, W. Ren, L. Gao, B. Liu, C. Jiang and H. M. Cheng, Synthesis of high-quality graphene with a pre-determined number of layers, *Carbon* 2009, 47, 493–9.

125. Y. Wu, B. Wang, Y. Ma, Y. Huang, N. Li, F. Zhang and Y. Chen, Efficient and largescale synthesis of few-layered graphene using an arc-discharge method and conductivity studies of the resulting films, *Nano Res.* 2010, 3, 661.

126. Z. Li, H. Zhu, D. Xie, K. Wang, A. Cao, J. Wei, X. Li, L. Fan and D. Wu, Flame synthesis of few-layered graphene/graphite films, *Chem. Commun.* 2011, 47, 3520.

127. P. Xu, Y. Yang, D. Qi, S. D. Barber, J. K. Schoelz, M. L. Ackerman, L. Bellaiche and P. M. Thibado, Electronic transition from graphite to graphene via controlled movement of the top layer with scanning tunneling microscopy, *Phys. Rev. B* 2012, 86, 085428.

128. R. Xu, L. J. Yin, J. B. Qiao, K. K. Bai, J. C. Nie and L. He, Direct probing of the stacking order and electronic spectrum of rhombohedral trilayer graphene with scanning tunneling microscopy, *Phys. Rev. B* 2015, 91, 035410.

129. D. Pierucci, H. Sediri, M. Hajlaoui, J. C. Girard, T. Brumme, M. Calandra, E. Velez-Fort, G. Patriarche, M. G. Silly, G. Ferro, V. Soulière, M. Marangolo,

F. Sirotti, F. Mauri and A. Ouerghi, Evidence for flat bands near the Fermi level in epitaxial rhombohedral multilayer graphene, *ACS Nano* 2015, 9, 5432.

130. Y. Que, W. Xiao, H. Chen, D. Wang, S. Du and H.-J. Gao, Stacking-dependent electronic property of trilayer graphene epitaxially grown on Ru (0001), *Appl. Phys. Lett.* 2015, 107, 263101.

131. K. F. Mak, J. Shan and T. F. Heinz, Electronic structure of few-layer graphene: Experimental demonstration of strong dependence on stacking sequence, *Phys. Rev. Lett.* 2010, 104, 176404.

132. R. R. Haering, Band structure of rhombohedral graphite, *Can. J. Phys.* 1958, 36, 352.

133. J. W. McClure, Electron energy band structure and electronic properties of rhombohedral graphite, *Carbon* 1969, 7, 425.

134. D. P. Arovas and F. Guinea, Stacking faults, bound states, and quantum Hall plateaus in crystalline graphite, *Phys. Rev. B* 2008, 78, 245416.

135. C. W. Chiu, Y. C. Huang, F. L. Shyu and M. F. Lin, Excitation spectra of ABC-stacked graphene superlattice. *Appl. Phys. Lett.* 2011, 98, 261920.

136. C. W. Chiu, Y. C. Huang, S. C. Chen, M. F. Lin and F. L. Shyu, Low-frequency electronic and optical properties of rhombohedral graphite, *Phys. Chem. Chem. Phys.* 2011, 13, 6036–42.

137. Y. P. Lin, C. Y. Lin, Y. H. Ho, T.N. Do and M. F. Lin, Magneto-optical properties of ABC-stacked trilayer graphene, *Phys. Chem. Chem. Phys.* 2015, 17, 15921.

138. C. H. Ho, C. P. Chang and M. F. Lin, Optical magnetoplasmons in rhombohedral graphite with a three-dimensional Dirac cone structure, *J. Phys. Condens. Matter* 2015, 27, 125602.

139. Y. Henni, H. P. O. Collado, K. Nogajewski, M. R. Molas, G. Usaj, C. A. Balseiro, M. Orlita, M. Potemski and C. Faugeras, Rhombohedral multilayer graphene: A magneto-Raman scattering study, *Nano Lett.* 2016, 16, 3710–6.

140. R. Saito, M. Fujita, G. Dresselhaus and M. S. Dresselhaus, Electronic structure of graphene tubules based on C_{60}, *Phys. Rev. B* 1992, 46, 1804.

141. C. L. Kane and E. J. Mele, Size, shape, and low energy electronic structure of carbon nanotubes, *Phys. Rev. Lett.* 1997, 78, 1932.

142. M. Ouyang, J. L. Huang, C. L. Cheung and C. M. Lieber, Energy gaps in "metallic" single-walled carbon nanotubes, *Science* 2001, 292, 702.

143. J. W. G. Wilder, L. C. Venema, A. G. Rinzler, R. E. Smalley and C. Dekker, Electronic structure of atomically resolved carbon nanotubes, *Nature* 1998, 391, 59–62.

144. T. W. Odom, J.-L. Huang, P. Kim and C. M. Lieber, Atomic structure and electronic properties of single-walled carbon nanotubes, *Nature* 1998, 391, 62–4.

145. M. F. Lin and K. W.-K. Shung, Magnetoconductance of carbon nanotubes, *Phys. Rev. B* 1995, 51, 7592.

146. H. Ajiki and T. J. Ando, Magnetic properties of carbon nanotubes, *J. Phys. Soc. Jap.* 1993, 62, 2470–80.

147. F. L. Shyu, C. P. Chang, R. B. Chen, C. W. Chiu and M. F. Lin, Magnetoelectronic and optical properties of carbon nanotubes, *Phys. Rev. B* 2003, 67, 045405.

148. S. Roche, G. Dresselhaus, M. S. Dresselhaus and R. Saito, Aharonov–Bohm spectral features and coherence lengths in carbon nanotubes, *Phys. Rev. B* 2000, 62, 16092.

149. S. Zaric, G. N. Ostojic, J. Kono, J. Shaver, V. C. Moore, M. S. Strano, R. H. Hauge, R. E. Smalley and X. Wei, Optical signatures of the Aharonov–Bohm phase in single-walled carbon nanotubes, *Science* 2004, 304, 1129.

150. N. Akima, Y. Iwasa, S. Brown, A. M. Barbour, J. B. Cao, J. L. Musfeldt, H. Matsui, N. Toyota, M. Shiraishi, H. Shimoda and O. Zhou, Strong anisotropy in the far-infrared absorption spectra of stretch-aligned single-walled carbon nanotubes. *Adv. Mater.* 2006, 18, 1166.

151. C. Jien, W. Qian and D. Hongjie, Electron transport in very clean, as-grown suspended carbon nanotubes. *Nat. Mater.* 2005, 4, 745–9.

152. C. Jien, W. Qian, R. Marco and D. Hongjie, Aharonov–Bohm interference and beating in single-walled carbon-nanotube interferometers, *Phys. Rev. Lett.* 2004, 93, 216803.

153. A. Bachtold, C. Strunk, J. P. Salvetat, J. M. Bonard, L. Forro, T. Nussbaumer and C. Schonenberger, Aharonov–Bohm oscillations in carbon nanotubes, *Nature* 1998, 397, 673–5.

154. H. Ajiki and T. Ando, Energy bands of carbon nanotubes in magnetic fields, *J. Phys. Soc. Jpn.* 1996, 65, 505.

155. J. W. Bai, X. F. Duan and Y. Huang, Rational fabrication of graphene nanoribbons using a nanowire etch mask, *Nano Lett.* 2009, 9, 2083–7.

156. L. C. Campos, V. R. Manfrinato, J. D. Sanchez-Yamagishi, J. Kong and P. Jarillo-Herrero, Anisotropic etching and nanoribbon formation in single-layer graphene, *Nano Lett.* 2009, 9, 2600–4.

157. A. G. Cano-Marquez, F. J. Rodriguez-Macias, J. Campos-Delgado, C. G. Espinosa-Gonzalez, F. Tristan-Lopez, D. Ramirez-Gonzalez, D. A. Cullen, D. J. Smith, M. Terrones and Y. I. Vega-Cantú, Graphene sheets and ribbons produced by lithium intercalation and exfoliation of carbon nanotubes, *Nano Lett.* 2009, 9, 1527–33.

158. F. Cataldo, G. Compagnini, G. Patane, O. Ursini, G. Angelini, P. R. Ribic, G. Margaritondo, A. Cricenti, G. Palleschi, F. Valentini, Graphene nanoribbons produced by the oxidative unzipping of single-wall carbon nanotubes, *Carbon* 2010, 48, 2596–602.

159. D. V. Kosynkin, A. L. Higginbotham, A. Sinitskii, J. R. Lomeda, A. Dimiev, B. K. Price and J. M. Tour, Longitudinal unzipping of carbon nanotubes to form graphene nanoribbons, *Nature* 2009, 458, 872–6.

160. Y. Z. Tan, B. Yang, K. Parvez, A. Narita, S. Osella, D. Beljonne, X. Feng and K. Müllen, Atomically precise edge chlorination of nanographenes and its application in graphene nanoribbons. *Nat. Comm.* 2013, 4, 2646.

161. J. M. Cai, P. Ruffieux, R. Jaafar, M. Bieri, T. Braun, S. Blankenburg, M. Muoth, A. P. Seitsonen, M. Saleh, X. Feng, K. Müllen and R. Fasel, Atomically precise bottom-up fabrication of graphene nanoribbons, *Nature* 2010, 466, 470–3.

162. J. Campos-Delgado, J. M. Romo-Herrera, X. T. Jia, D. A. Cullen, H. Muramatsu, Y. A. Kim, T. Hayashi, Z. Ren, D. J. Smith, Y. Okuno, T. Ohba, H. Kanoh, K. Kaneko, M. Endo, H. Terrones, M. S. Dresselhaus and M. Terrones, Bulk production of a new form of sp^2 carbon: Crystalline graphene nanoribbons, *Nano Lett.* 2008, 8, 2773–8.

163. P. Ruffieux, J. Cai, N. C. Plumb, L. Patthey, D. Prezzi, A. Ferretti and R. Fasel, Electronic structure of atomically precise graphene nanoribbons, *ACS Nano* 2012, 6, 6930–5.

164. Y. Sugiyama, O. Kubo, R. Omura, M. Shigehara, H. Tabata, N. Mori and M. Katayama, Spectroscopic study of graphene nanoribbons formed by crystallographic etching of highly oriented pyrolytic graphite. *Appl. Phys. Lett.* 2014, 105, 123116.

165. H. Huang, D. Wei, J. Sun, S. L. Wong, Y. P. Feng, A. C. Neto and A. T. S. Wee, Spatially resolved electronic structures of atomically precise armchair graphene nanoribbons, *Sci. Rep.* 2012, 2, 983.

166. H. Söde, L. Talirz, O. Gröning, C. A. Pignedoli, R. Berger, X. Feng and P. Ruffieux, Electronic band dispersion of graphene nanoribbons via Fourier-transformed scanning tunneling spectroscopy, *Phys. Rev. B* 2015, 91, 045429.

167. Y. C. Chen, D. G. De Oteyza, Z. Pedramrazi, C. Chen, F. R. Fischer and M. F. Crommie, Tuning the band gap of graphene nanoribbons synthesized from molecular precursors, *ACS Nano* 2013, 7, 6123–8.

168. H. C. Chung, M. H. Lee, C. P. Chang and M. F. Lin, Exploration of edge-dependent optical selection rules for graphene nanoribbons, *Opt. Exp.* 2011, 19, 23350.

169. V. A. Saroka, M. V. Shuba and M. E. Portnoi, Optical selection rules of zigzag graphene nanoribbons, *Phys. Rev. B* 2017, 95, 155438.

170. K. I. Sasaki, K. Kato, Y. Tokura, K. Oguri and T. Sogawa, Theory of optical transitions in graphene nanoribbons, *Phys. Rev. B* 2011, 84, 085458.

171. N. M. R. Peres, A. H. Castro Neto and F. Guinea, Dirac fermion confinement in graphene, *Phys. Rev B* 2006, 73, 241403–7(R).

172. Y. C. Huang, C. P. Chang and M. F. Lin, Magnetic and quantum confinement effects on electronic and optical properties of graphene ribbons, *Nanotechnology* 2007, 18, 495401–9.

173. Y. C. Huang, M. F. Lin and C. P. Chang, Landau levels and magnetooptical properties of graphene ribbons, *J. Appl. Phys.* 2008, 103, 073709–16.

174. Y. C. Ou, J. K. Sheu, Y. H. Chiu, R. B. Chen and M. F. Lin, Influence of modulated fields on the Landau level properties of graphene, *Phys. Rev. B* 2011, 83, 195405.

175. Y. H. Chiu, Y. C. Ou, Y. Y. Liao and M. F. Lin, Optical-absorption spectra of single-layer graphene in a periodic magnetic field, *J. Vac. Sci. Technol. B* 2010, 28, 386–90.

176. Y. C. Ou, Y. H. Chiu, P. H. Yang and M. F. Lin, The selection rule of graphene in a composite magnetic field, *Opt. Exp.* 2014, 22, 7473–91.

177. J. Y. Wu, S. C. Chen, G. Gumbs and M. F. Lin, Feature-rich electronic excitations in external fields of 2D silicone, *Phys. Rev. B* 2016, 94, 205427.

178. J. Y. Wu, S. C. Chen and M. F. Lin, Temperature-dependent Coulomb excitations in silicone, *New J. Phys.* 2014, 16, 125002.

179. J. Y. Wu, C. Y. Lin, G. Gumbs and M. F. Lin, The effect of perpendicular electric field on temperature-induced plasmon excitations for intrinsic silicene, *RSC Adv.* 2015, 5, 51912–8.

180. P. H. Shih, Y. H. Chiu, J. Y. Wu, F. L. Shyu and M. F. Lin, Coulomb excitations of monolayer germanene. *Sci. Rep.* 2017, 7, 40600.

181. S. C. Chen, C. L. Wu, J. Y. Wu and M. F. Lin, Magnetic quantization of sp^3 bonding in monolayer gray tin, *Phys. Rev. B* 2016, 94, 045410.

182. G. Binnig and H. Rohrer, Scanning tunneling microscopy, *IBM J. Res. Dev.* 1986, 30, 355–69.

183. B. Zha, M. Q. Dong, X. R. Miao, S. Peng, Y. C. Wu, K. Miao, Y. Hu and W. Deng, Cooperation and competition between halogen bonding and van der Waals forces in supramolecular engineering at the aliphatic hydrocarbon/graphite interface: Position and number of bromine group effects, *Nanoscale* 2017, 9, 237–50.

184. D. Yildiz, H. Ş. Ş. O. Gürlseren and O. Gürlü, Apparent corrugation variations on moiré patterns on highly oriented pyrolytic graphite, *Mater. Today Comm.* 2016, 8, 72–8.

185. I. Miccoli, J. Aprojanz, J. Baringhaus, T. Lichtenstein, L. A. Galves, J. M. J. Lopes and C. Tegenkamp, Quasi-free-standing bilayer graphene nanoribbons probed by electronic transport, *Appl. Phys. Lett.* 2017, 110, 051601.

186. J. J. Song, H. J. Zhang, Y. L. Cai, Y. X. Zhang, S. N. Bao and P. M. He, Bottom-up fabrication of graphene nanostructures on Ru(eqn_0002.eps), *Nanotechnology* 2016, 27, 055602.

187. K. A. Simonov, N. A. Vinogradov, A. S. Vinogradov, A. V. Generalov, E. M. Zagrebina, G. I. Svirskiy, A. A. Cafolla, T. Carpy, J. P. Cunniffe, T. Taketsugu, A. Lyalin, N. Mårtensson and A. B. Preobrajenski, From graphene nanoribbons on Cu(111) to nanographene on Cu(110): Critical role of substrate structure in the bottom-up fabrication strategy, *ACS Nano* 2015, 9, 8997–9011.

188. J. Z. Liu, B. W. Li, Y. Z. Tan, A. Giannakopoulos, C. Sanchez-Sanchez, D. Beljonne, P. Ruffieux, R. Fasel, X. Feng and K. Müllen, Toward cove-edged low band gap graphene nanoribbons, *J. Am. Chem. Soc.* 2015, 137, 6097–103.

189. B. V. Andryushechkin, V. M. Shevlyuga, T. V. Pavlova, G. M. Zhidomirov and K. N. Eltsov, Adsorption of O^2 on Ag(111): Evidence of local oxide formation, *Phys. Rev. Lett.* 2016, 117, 056101.

190. H. González-Herrero, P. Pou, J. Lobo-Checa, D. Fernández-Torre, F. Craes, A. J. Martínez-Galera, M. M. Ugeda, M. Corso, J. E. Ortega, J. M. Gómez-Rodríguez, R. Pérez and I. Brihuega, Graphene tunable transparency to tunneling electrons: A direct tool to measure the local coupling, *ACS Nano* 2016, 10, 5131–44.

191. O. E. Dagdeviren, C. Zhou, K. Zou, G. H. Simon, S. D. Albright, S. Mandal, M. D. Morales-Acosta, X. Zhu, S. Ismail-Beigi, F. J. Walker, C. H. Ahn, U. D. Schwarz and E. I. Altman, Length scale and dimensionality of defects in epitaxial SnTe topological crystalline insulator films. *Adv. Mater. Interfaces* 2017, 4, 1601011.

192. L. R. Merte, Y. H. Bai, H. Zeuthen, G. W. Peng, L. Lammich, F. Besenbacher, M. Mavrikakisb and S. Wendt, Identification of O-rich structures on platinum(111)-supported ultrathin iron oxide films, *Surf. Sci.* 2016, 652, 261–8.

193. M. Setvin, M. Wagner, M. Schmid, G. S. Parkinson and U. Diebold, Surface point defects on bulk oxides: Atomically-resolved scanning probe microscopy, *Chem. Soc. Rev.* 2017, 46, 1772–84.

194. S. M. Hollen, S. J. Tjung, K. R. Mattioli, G. A. Gambrel, N. M. Santagata, E. Johnston-Halperin, et al., Native defects in ultra-high vacuum grown graphene islands on Cu(111), *J. Phys. Condens. Matter* 2016, 28, 034003.

195. D. T. Pierce, Spin-polarized electron microscopy, *Phys. Scr.* 1988, 38, 291.

196. R. Wiesendanger, H.-J. Güntherodt, G. Güntherodt, R. Gambino and R. Ruf, Observation of vacuum tunneling of spin-polarized electrons with the scanning tunneling microscope, *Phys. Rev. Lett.* 1990, 65, 247.

197. Z. B. Liu, Z. Y. Fei, C. Xu, Y. X. Jiang, X. L. Ma, H. M. Chenga and W. Ren, Phase transition and in situ construction of lateral heterostructure of 2D superconducting α/β Mo^2C with sharp interface by electron beam irradiation, *Nanoscale* 2017, 9 7501–7.

198. D. D. Su, Y. Y. Zhang, Z. J. Wang, Q. J. Wan and N. J. Yang, Decoration of graphene nano platelets with gold nanoparticles for voltammetry of 4-nonylphenor, *Carbon* 2017, 117, 313–21.

199. K. Miyazawa, M. Watkins, A. L. Shluger and T. Fukuma, Influence of ions on two-dimensional and three-dimensional atomic force microscopy at fluorite-water interfaces, *Nanotechnology* 2017, 28, 245701.

200. M. Behzadirad, M. Nami, A. K. Rishinaramagalam, D. F. Feezell and T. Busani, GaN nanowire tips for nanoscale atomic force microscopy, *Nanotechnology* 2017, 28, 20LT01.

201. Z. W. Dai, W. C. Jin, M. Grady, J. T. Sadowski, J. I. Dadap, R. M. Osgood and K. Pohl, Surface structure of bulk 2H-MoS^2(0001) and exfoliated suspended monolayer MoS^2: A selected area low energy electron diffraction study, *Surf. Sci.* 2017, 660, 16–21.

202. K. Kawahara, T. Shirasawa, C. L. Lin, R. Nagao, N. Tsukahara, T. Takahashi, R. Arafune, M. Kawai and N. Takagi, Atomic structure of "multilayer silicene" grown on Ag(111): Dynamical low energy electron diffraction analysis, *Surf. Sci.* 2016, 651, 70–5.

203. C. J. Chen, *Introduction to Scanning Tunneling Microscopy*, Oxford: Oxford University Press 1993, ISBN 0-19-507150-6.

204. C. P. Leon, H. Drees, S. M. Wippermann, M. Marz and R. Hoffmann-Vogel, Atomically resolved scanning force studies of vicinal Si(111), *Phys. Rev. B* 2017, 95, 245412.

205. V. Jelic, K. Iwaszczuk, P. H. Nguyen, C. Rathje, G. J. Hornig, H. M. Sharum, J. R. Hoffman, M. R. Freeman and F. A. Hegmann, Ultrafast terahertz control of extreme tunnel currents through single atoms on a silicon surface. *Nat. Phys.* 2017, 13, 591–8.

206. P. Mondelli, B. Gupta, M. G. Betti, C. Mariani, J. L. Duffin and N. Motta, High quality epitaxial graphene by hydrogenetching of 3C-SiC(111) thin-film on Si(111), *Nanotechnology* 2017, 28, 115601.

207. J. W. Wilder, L. C. Venema, A. G. Rinzler, R. E. Smalley and C. Dekker, Electronic structure of atomically resolved carbon nanotubes, *Nature* 1998, 391, 59.

208. T. W. Odom, J.-L. Huang, P. Kim and C. M. Lieber, Atomic structure and electronic properties of single-walled carbon nanotubes, *Nature* 1998, 391, 62.

209. S. Tanaka, M. Matsunami and S. Kimura, An investigation of electron–phonon coupling via phonon dispersion measurements in graphite using angle-resolved photoelectron spectroscopy, *Sci. Rep.* 2013, 3, 3031.

210. M. S. Epstein and T. C. Rains, Evaluation of a xenon-mercury arc lamp for background correction in atomic-absorption spectrometry, *Anal. Chem.* 1976, 48, 528–31.

211. G. L. Stamm, R. L. Denningh and A. G. Rockman, Some operating characteristics of a xenon and a xenon-mercury short-arc lamp immersed in water, Report of NRL progress FEB, 1969, 33.

212. L. G. Ferguson and F. Dogan, Spectrally selective, matched emitters for thermophotovoltaic energy conversion processed by tape casting, *J. Mater. Sci.* 2001, 36, 137–46.

213. H. L. Zhen, N. Li, D. Y. Xiong, X. C. Zhou, W. Lu and H. C. Liu, Fabrication and investigation of an upconversion quantum-well infrared photodetector integrated with a light-emitting diode, *Chin. Phys. Lett.* 2005, 22, 1806–8.

214. J. E. Stewart and J. C. Richmond, Infrared emission spectrum of silicon carbide heating elements, *J. Res. Natl. Bur. Stand.* 1957, 59, 405–9.

215. S. H. Lu, W. C. Liu and J. P. Liu, High-axial-resolution, full-field optical coherence microscopy using tungsten halogen lamp and liquid-crystal-based achromatic phase shifter, *Appl. Opt.* 2015, 54, 4447–52.

216. J. F. Wei, X. Y. Hu, L. Q. Sun, K. Zhang and Y. Chang, Technology for radiation efficiency measurement of high-power halogen tungsten lamp used in calibration of high-energy laser energy meter, *Appl. Opt.* 2015, 54, 2289–95.

217. H. Keppler, L. S. Dubrovinsky, O. Narygina and I. Kantor, Optical absorption and radiative thermal conductivity of silicate perovskite to gigapascals, *Science* 2008, 322, 1529–32.

218. S. Albert, K. K. Albert, P. Lerch and M. Quack, Synchrotron-based highest resolution Fourier transform infrared spectroscopy of naphthalene ($C_{10}H_8$) and indole (C_8H_7N) and its application to astrophysical problems, *Faraday Discuss* 2011, 150, 71–99.

219. G. A. Gasparian and H. Lucht, Indium gallium arsenide NIR photodiode array spectroscopy, *Spectroscopy* 2000, 15, 16.

220. R. E. Dessy, W. G. Nunn, C. A. Itus and W. R. Reynolds, Linear photodiode array spectrometers as detector systems in automated liquid chromatographs, *J. Chromatogr. Sci.* 1976, 14, 195–200.

221. C. Clementi, W. Nowik, A. Romani, D. Cardon, M. Trojanowicz, A. Davantes, and P. Chaminade, Towards a semiquantitative non invasive characterisation of Tyrian purple dye composition: Convergence of UV-visible reflectance spectroscopy and fast-high temperature-high performance liquid chromatography with photodiode array detection, *Anal. Chim. Acta.* 2016, 926, 17–27.

222. J. B. Johnson, G. Edwards and M. Mendenhall, Low-cost, high-performance array detector for spectroscopy based on a charge-coupled photodiode, *Rev. Sci. Instrum.* 1994, 65, 1782–3.

223. I. Johnson, Z. Sadygov, O. Bunk, A. Menzel, F. Pfeiffer and D. Renker, A Geiger-mode avalanche photodiode array for X-ray photon correlation spectroscopy, *J. Synchrotron Radiat.* 2009, 16, 105–9.

224. V. B. Podobedov, C. C. Miller and M. E. Nadal, Performance of the NIST goniocolorimeter with a broad-band source and multichannel charged coupled device based spectrometer, *Rev. Sci. Instrum.* 2012, 83, 093108.

225. L. F. Lastras-Martinez, R. Castro-Garcia, R. E. Balderas-Navarro and A. Lastras-Martinez, Microreflectance difference spectrometer based on a charge coupled device camera: Surface distribution of polishing-related linear defect density in GaAs (001), *Appl. Opt.* 2009, 48, 5713–7.

226. C. Fourment, N. Arazam, C. Bonte, T. Caillaud, D. Descamps, F. Dorchies, M. Harmand, S. Hulin, S. Petit and J. J. Santos, Broadband, high dynamics and high resolution charge coupled device-based spectrometer in dynamic mode for multi-keV repetitive x-ray sources, *Rev. Sci. Instrum.* 2009, 80, 083505.

227. S. Awaji, K. Watanabe, H. Oguro, H. Miyazaki, S. Hanai, T. Tosaka and S. Ioka, First performance test of a 25 T cryogen-free superconducting magnet, *Supercond. Sci. Technol.* 2017, 30, 065001.

228. M. Takahashi, S. Iwai, H. Miyazaki, T. Tosaka, K. Tasaki, S. Hanai, S. Ioka, H. Takigami, K. Watanabe, S. Awaji, H. Oguro and Y. Tsuchiya, Design and test results of a cryogenic cooling system for a 25-T cryogen-free superconducting magnet, *IEEE Trans. Appl. Supercond.* 2017, 27, 4603805.

229. R. Sakakura, Y. H. Matsuda, M. Tokunaga, E. Kojima and S. Takeyama, Application of an electro-magnetic induction technique for the magnetization up to 100 T in a vertical single-turn coil system, *J. Low Temp. Phys.* 2010, 159, 297–301.

230. C. P. Chang, Exact solution of the spectrum and magneto-optics of multilayer hexagonal graphene, *J. Appl. Phys.* 2011, 110, 013725.
231. C. W. Chiu, S. H. Lee, S. C. Chen and M. F. Lin, Absorption spectra of AA-stacked graphite, *New J. Phys.* 2010, 12, 083060.
232. K. Sugawara, T. Sato, S. Souma, T. Takahashi and H. Suematsu, Fermi surface and edge-localized states in graphite studied by high-resolution angle-resolved photoemission spectroscopy, *Phys. Rev. B* 2006, 73, 045124.
233. D. A. Siegel, W. Regan, A. V. Fedorov, A. Zettl and A. Lanzara, Charge-carrier screening in single-layer graphene, *Phys. Rev. Lett.* 2013, 110, 146802.
234. A. Bostwick, T. Ohta, T. Seyller, K. Horn and E. Rotenberg, Quasiparticle dynamics in graphene, *Nat. Phys.* 2007, 3, 36–40.
235. A. Luican, Guohong Li, A. Reina, J. Kong, R. R. Nair, K. S. Novoselov, A. K. Geim and E. Y. Andrei, Single-layer behavior and its breakdown in twisted graphene layers, *Phys. Rev. Lett.* 2011, 106, 126802.
236. G. Li, A. Luican, J. L. Dos Santos, A. C. Neto, A. Reina, J. Kong and E. Y. Andrei, Observation of van Hove singularities in twisted graphene layers, *Nat. Phys.* 2010, 6, 109–13.
237. V. Cherkez, G. T. de Laissardiere, P. Mallet and J.-Y. Veuillen, Van Hove singularities in doped twisted graphene bilayers studied by scanning tunneling spectroscopy, *Phys. Rev. B* 2015, 91, 155428.
238. M. Yankowitz, F. Wang, C. N. Lau and B. J. LeRoy, Local spectroscopy of the electrically tunable band gap in trilayer graphene, *Phys. Rev. B* 2013, 87, 165102.
239. F. Guinea, M. Katsnelson and M. Vozmediano, Midgap states and charge inhomogeneities in corrugated graphene, *Phys. Rev. B* 2008, 77, 075422.
240. F. Guinea, B. Horovitz and P. Le Doussal, Gauge field induced by ripples in graphene, *Phys. Rev. B* 2008, 77, 205421.
241. C. W. Chiu, S. C. Chen, Y. C. Huang, F. L. Shyu and M. F. Lin, Critical optical properties of AA-stacked multilayer graphenes, *Appl. Phys. Lett.* 2013, 103, 041907.
242. A. Narita, X. Feng, Y. Hernandez, S. A. Jensen, M. Bonn, H. Yang, I. A. Verzhbitskiy, C. Casiraghi, M. R. Hansen, A. H. R. Koch, G. Fytas, O. Ivasenko, B. Li, K. S. Mali, T. Balandina, S. Mahesh, S. De Feyter and K. Müllen, Synthesis of structurally well-defined and liquid-phase-processable graphene nanoribbons, *Nat. Chem.* 2014, 6, 126–32.
243. L. Degiorgi, E. J. Nicol, O. Klein, G. Gruner, P. Wachter, S.-M. Huang, J. Wiley and R. B. Kaner, Optical properties of the alkali-metal-doped superconducting fullerenes: K^3C^{60} and Rb^3C^{60}, *Phys. Rev. B* 1994, 49, 7012.
244. K. Harigaya and S. Abe, Optical-absorption spectra in fullerenes C^{60} and C^{70}: Effects of Coulomb interactions, lattice fluctuations, and anisotropy, *Phys. Rev. B* 1994, 49, 16746.
245. I. A. Luk'yanchuk and Y. Kopelevich, Phase analysis of quantum oscillations in graphite, *Phys. Rev. Lett.* 2004, 93, 166402.
246. N. Ubrig, P. Plochocka, P. Kossacki, M. Orlita, D. K. Maude, O. Portugall and G. L. J. A. Rikken, High-field magnetotransmission investigation of natural graphite, *Phys. Rev. B* 2011, 83, 073401.
247. P. A. Obraztsov, G. M. Mikheev, S. V. Garnov, A. N. Obraztsov and Y. P. Svirko, Polarization-sensitive photoresponse of nanographite, *Appl. Phys. Lett.* 2011, 98, 091903.

248. J. M. Zhang and P. C. Eklun, Optical transmission of graphite and potassium graphite intercalation compounds, *J. Mater. Res.* 1987, 2, 858–63.

249. E. Jung, S. Lee, S. Roh, X. Meng, S. Tongay, J. Kang, T. Park and J. Hwang, Optical properties of NbCl5 and ZnMg intercalated graphite compounds, *J. Phys. D: Appl. Phys.* 2014, 47, 485304.

250. D. S. Kim, H. Kwon, A. Y. Nikitin, S. Ahn, L. Martin-Moreno, F. J. Garcia-Vidal, S. Ryu, H. Min and Z. H. Kim, Stacking structures of few-layer graphene revealed by phase-sensitive infrared nanoscopy, *ACS Nano* 2015, 9, 6765–73.

251. M. Koshino and E. McCann, Landau level spectra and the quantum Hall effect of multilayer graphene, *Phys. Rev. B* 2011, 83, 165443.

252. L.-C. Tung, P. Cadden-Zimansky, J. Qi, Z. Jiang and D. Smirnov, Measurement of graphite tight-binding parameters using high-field magnetoreflectance, *Phys. Rev. B* 2011, 84, 153405.

253. P. Kossacki, C. Faugeras, M. Kuhne, M. Orlita, A. A. L. Nicolet and J. M. Schneider, Electronic excitations and electron-phonon coupling in bulk graphite through Raman scattering in high magnetic fields, *Phys. Rev. B* 2011, 84, 235138.

254. Y. Kim, Y. Ma, A. Imambekov, N. G. Kalugin, A. Lombardo, A. C. Ferrari, J. Kono and D. Smirnov, Magnetophonon resonance in graphite: High-field Raman measurements and electron-phonon coupling contributions. *Phys. Rev. B* 2012, 85, 121403(R).

255. P. Plochocka, P. Y. Solane, R. J. Nicholas, J. M. Schneider, B. A. Piot, D. K. Maude, et al., Origin of electron-hole asymmetry in graphite and graphene. *Phys. Rev. B* 2012, 85, 245410.

256. R. J. Nicholas, P. Y. Solane and O. Portugall, Ultrahigh magnetic field study of layer split bands in graphite, *Phys. Rev. Lett.* 2013, 111, 096802.

257. E. A. Henriksen, Z. Jiang, L.-C. Tung, M. E. Schwartz, M. Takita, Y.-J. Wang, P. Kim, and H. L. Stormer, Cyclotron resonance in bilayer graphene, *Phys. Rev. Lett.* 2008, 100, 087403.

258. J. M. Pereira Jr., F. M. Peeters and P. Vasilopoulos, Landau levels and oscillator strength in a biased bilayer of graphene, *Phys. Rev. B* 2007, 76, 115419.

259. A. B. Kuzmenko, E. van Heumen, D. van der Marel, P. Lerch, P. Blake, K. S. Novoselov and A. K. Geim, Infrared spectroscopy of electronic bands in bilayer graphene, *Phys. Rev. B* 2009, 79, 115441.

260. L. M. Zhang, Z. Q. Li, D. N. Basov and M. M. Fogler, Determination of the electronic structure of bilayer graphene from infrared spectroscopy, *Phys. Rev. B* 2008, 78, 235408.

261. M. L. Sadowski, G. Martinez, M. Potemski, C. Berger and W. A. de Heer, Landau level spectroscopy of ultrathin graphite layers, *Phys. Rev. Lett.* 2006, 97, 266405.

262. S. Berciaud, M. Potemski and C. Faugeras, Probing electronic excitations in mono- to pentalayer graphene by micro magneto-Raman spectroscopy, *Nano Lett.* 2014, 14, 4548–53.

263. C. Faugeras, M. Amado, P. Kossacki, M. Orlita, M. Kuhne, A. A. L. Nicolet, Yu. I. Latyshev and M. Potemski, Magneto-Raman scattering of graphene on graphite: Electronic and phonon excitations, *Phys. Rev. Lett.* 2011, 107, 036807.

264. S. Yuan, R. Roldan and M. I. Katsnelson, Landau level spectrum of ABA- and ABC-stacked trilayer graphene, *Phys. Rev. B* 2011, 84, 125455.

265. J. W. McClure, Electron energy band structure and electronic properties of rhombohedral graphite, *Carbon* 1969, 7, 425.

266. M. Taut, K. Koepernik and M. Richter, Electronic structure of stacking faults in rhombohedral graphite, *Phys. Rev. B* 2014, 90, 085312.

267. C. H. Ho, C. P. Chang and M. F. Lin, Evolution and dimensional crossover from the bulk subbands in ABC-stacked graphene to a three-dimensional Dirac cone structure in rhombohedral graphite, *Phys. Rev. B* 2016, 93, 075437.

268. R. Xiao, F. Tasnadi, K. Koepernik, J. W. F. Venderbos, M. Richter and M. Taut, Density functional investigation of rhombohedral stacks of graphene: Topological surface states, nonlinear dielectric response, and bulk limit, *Phys. Rev. B* 2011, 84, 165404.

269. C. L. Lu, C. P. Chang, Y.C. Huang, J. H. Ho, C. C. Hwang and M. F. Lin, Electronic properties of AA- and ABC-stacked few-layer graphites, *J. Phys. Soc. Jpn.* 2007, 76, 024701.

270. C. W. Chiu, Y. C. Huang, F. L. Shyu and M. F. Lin, Optical absorption spectra in ABC-stacked graphene superlattice, *Synth. Met.* 2012, 162, 800.

271. C. Y. Lin, T. N. Do, Y. K. Huang and M. F. Lin, Optical properties of graphene in magnetic and electric fields, arXiv:1603.02797

272. C. H. Lui, Z. Li, K. F. Mak, E. Cappelluti and T. F. Heinz, Observation of an electrically tunable band gap in trilayer graphene, *Nat. Phys.* 2011, 7, 944–7.

273. C. H. Ho, Y. H. Ho, Y. Y. Liao, Y. H. Chiu, C. P. Chang and M. F. Lin, Diagonalization of Landau level spectra in rhombohedral graphite, *J. Phys. Soc. Jpn.* 2012, 81, 024701.

274. T. T. Heikkila and G. E. Volovik, Dimensional crossover in topological matter: Evolution of the multiple Dirac point in the layered system to the flat band on the surface, *JETP Lett.* 2011, 93, 59.

275. J. Jiang, R. Saito, A. Grüneis, G. Dresselhaus and M. S. Dresselhaus, Optical absorption matrix elements in single-wall carbon nanotubes, *Carbon* 2004, 42, 3169–73.

276. M. F. Lin, Optical spectra of single-wall carbon nanotube bundles, *Phys. Rev. B* 2000, 62, 13153–9.

277. M. Y. Sfeir, T. Beetz, F. Wang, L. Huang, X. M. H. Huang, M. Huang, J. Hone, S. O'Brien, J. A. Misewich, T. F. Heinz, L. Wu, Y. Zhu and L. E. Brus, Optical spectroscopy of individual single-walled carbon nanotubes of defined chiral structure, *Science* 2006, 312, 554–6.

278. M. Y. Sfeir, F. Wang, L. Huang, C. C. Chuang, J. Hone, S. P. O'Brien, T. F. Heinz, L. E. Brus, Probing electronic transitions in individual carbon nanotubes by Rayleigh scattering, *Science* 2004, 306, 1540–3.

279. C. Y. Lin, S. C. Chen, J. Y. Wu and M. F. Lin, Curvature effects on magnetoelectronic properties of nanographene ribbons, *J. Phys. Soc. Jpn.* 2012, 81, 064719.

280. M. F. Lin and K. W.-K. Shung, Magnetization of graphene tubules, *Phys. Rev. B* 1995, 52, 8423.

281. J. C. Charlier and P. Lambin, Electronic structure of carbon nanotubes with chiral symmetry, *Phys. Rev. B* 1998, 57, 15037–9.

282. M. S. Dresselhaus and G. Dresselhaus, Intercalation in layered materials, *Adv Phys.* 1981, 30, 139.

283. A. L. Woodhead, M. L. de Souza and J. S. Church, An investigation into the surface heterogeneity of nitric acid oxidized carbon fiber, *Appl. Surf. Sci.* 2016, 401, 79–88.

284. X. Zhang, X. Li, G. Yuan, Z. Dong, G. Mac and B. Rand, Large diameter pitch-based graphite fiber reinforced unidirectional carbon/carbon composites with high thermal conductivity densified by chemical vapor infiltration, *Carbon* 2017, 114, 59–69.

285. Y. Rew, A. Baranikumar, A. V. Tamashausky, S. E. Tawil and P. Park, Electrical and mechanical properties of asphaltic composites containing carbon based fillers, *Constr. Build. Mater.* 2017, 135, 394–404.

286. G. H. Li, X. J. Tian, X. W. Xu, C. Zhou, J. Y. Wu, Q. Li, L. Q. Zhang, F. Yang and Y. F. Li, Fabrication of robust and highly thermally conductive nanofibrillated cellulose/graphite nanoplatelets composite papers, *Compos. Sci. Technol.* 2017, 138, 179–85.

287. H. Fan, L. Qi and H. Wang, Hexafluorophosphate anion intercalation into graphite electrode from methyl propionate, *Solid State Ionics* 2017, 300, 169–74.

288. R. Matsumoto and Y. Okabe, Highly electrically conductive and air-stable metal chloride ternary graphite intercalation compounds with $AlCl_3$-$FeCl_3$ and $AlCl_3$-$CuCl_2$ prepared from flexible graphite sheets, *Synth. Met.* 2016, 222, 351–5.

289. X. Bie, K. Kubota, T. Hosaka, K. Chihara and S. Komaba, A novel K-ion battery: Hexacyanoferrate(II)/graphite cell, *J. Mater. Chem. A* 2017, 5, 4325–30.

290. Z. Tian, P. Yu, S. E. Lowe, A. G. Pandolfoa, T. R. Gengenbachc, K. M. Nairna, J. Song, X. Wang, Y. L. Zhong and D. Li, Facile electrochemical approach for the production of graphite oxide with tunable chemistry, *Carbon* 2017, 112, 185–91.

291. F. Cheng, G. J. Wang, Z. X. Sun, Y. Yu, F. Huang, C. L. Gong, H. Liu, G. Zheng, C. Qin and S. Wen, Carbon-coated SiO/ZrO_2 composites as anode materials for lithium-ion batteries, *Ceram. Int.* 2017, 43, 4309–13.

292. S. Kesavan, D. R. Kumar, Y. R. Lee and J. J. Shim, Determination of tetracycline in the presence of major interference in human urine samples using polymelamine/electrochemically reduced graphene oxide modified electrode, *Sens. Actuators B Chem.* 2017, 241, 455–65.

293. N. Vishnu and A. S. Kumar, Development of Prussian Blue and Fe(bpy)32+ hybrid modified pencil graphite electrodes utilizing its intrinsic iron for electroanalytical applications, *J. Electroanal. Chem.* 2017, 786, 145–53.

294. C. Rajkumar, B. Thirumalraj and S. M. Chen, A simple preparation of graphite/gelatin composite for electrochemical detection of dopamine, *J. Colloid Interface Sci.* 2017, 487, 149–55.

295. A. Ravalli, C. Rossi and G. Marrazza, Bio-inspired fish robot based on chemical sensors, *Sens. Actuators B Chem.* 2017, 239, 325–329.

296. M. Dervisevic, M. Senel, T. Sagir and S. Isik, Highly sensitive detection of cancer cells with an electrochemical cytosensor based on boronic acid functional polythiophene, *Biosens. Bioelectron.* 2016, 90, 6–12.

297. Z. J. Deng, H. Y. Long, Q. P. Wei, Z. M. Yu, B. Zhou, Y. J. Wang, L. Zhang, S, S. Li, L. Ma, Y. Xie and J. Min, High-performance non-enzymatic glucose sensor based on nickel-microcrystalline graphite-boron doped diamond complex electrode, *Sens. Actuators B Chem.* 2017, 242, 825–34.

298. L. Wang, Q. R. Xiong, F. Xiao and H. W. Duan, 2D nanomaterials based electrochemical biosensors for cancer diagnosis, *Biosens. Bioelectron.* 2017, 89, 136–51.

299. T. E. Weller, M. Ellerby, S. S. Saxena, R. P. Smith and N. T. Skipper, Superconductivity in the intercalated graphite compounds C_6Yb and C_6Ca, *Nat. Phys.* 2005, 1, 39–41.

300. G. Csanyi, P. B. Littlewood, A. H. Nevidomskyy, C. J. Pickard and B. D. Simons, The role of the interlayer state in the electronic structure of superconducting graphite intercalated compounds, *Nat. Phys.* 2005, 1, 42–5.
301. D. Frattini, G. Accardo, C. Ferone and R. Cioffi, Fabrication and characterization of graphite-cement composites for microbial fuel cells applications, *Mater. Res. Bull.* 2016, 88, 188–99.
302. J. Li, C. Liu, Q. Liao, X. Zhu and D. Ye, Improved performance of a tubular microbial fuel cell with a composite anode of graphite fiber brush and graphite granules, *Int. J. Hydrogen Energy* 2013, 38, 15723.
303. Y. Liu, F. Harnisch, U. Schröder, K. Fricke, V. Climent and J. M. Feliu, The study of electrochemically active microbial biofilms on different, carbon-based anode materials in microbial fuel cells, *Biosens. Bioelectron.* 2010, 25, 2167–71.
304. K. Said, A. I. Ayesh, N. N. Qamhieh, F. Awwad, S. T. Mahmoud and S. Hisaindee, Fabrication and characterization of graphite oxide-nanoparticle composite based field effect transistors for non-enzymatic glucose sensor applications, *J. Alloys Compd.* 2017, 694, 1061–6.
305. B. Standley, A. Mendez, E. Schmidgall and M. Bockrath, Graphene-graphite oxide field-effect transistors, *Nano Lett.* 2012, 12, 1165–9.
306. M. Burghard, H. Klauk and K. Kern, Carbon-based field-effect transistors for nanoelectronics, *Adv. Mater.* 2009, 21, 2586–600.
307. S. Bae, H. Kim, Y. Lee, X. Xu, J. S. Park, Y. Zheng, J. Balakrishnan, T. Lei, H. R. Kim, Y. I. Song, Y.-J. Kim, K. S. Kim, B. Özyilmaz, J.-H. Ahn, B. H. Hong and S. Iijima, Roll-to-roll production of 30-inch graphene films for transparent electrodes, *Nat Nanotechnol.* 2010, 5, 574–8.
308. K. O-Rak, S. Ummartyotin, M. Sain and H. Manuspiy, Covalently grafted carbon nanotube on bacterial cellulose composite for flexible touch screen application, *Mat. Lett.* 2013, 107, 247–50.
309. P. Matyba, H. Yamaguchi, G. Eda, M. Chhowalla, I. Edman and N. D. Robinson, Graphene and mobile ions: The key to all-plastic, solution-processed light-emitting devices, *ACS Nano* 2010, 4, 637–42.
310. Z. K. Zhang, J. H. Du, D. D. Zhang, H. D. Sun, L. C. Yin, L. P. Ma, J. S. Chen, D. G. Ma, H.-M. Cheng and W. C. Ren, Rosin-enabled ultraclean and damage-free transfer of graphene for large-area flexible organic light-emitting diodes, *Nat. Commun.* 2017, 8, 14560.
311. Z. Yu, L. Hu, Z. Liu, M. Sun, M. Wang, G. Gruner and Q. Pei, Fully bendable polymer light emitting devices with carbon nanotubes as cathode and anode, *Appl. Phys. Lett.* 2009, 95, 203304.
312. K. Lee, Z. Wu, Z. Chen, F. Ren, S. J. Pearton and A. G. Rinzler, Single wall carbon nanotubes for p-type ohmic contacts to GaN light-emitting diodes, *Nano Lett.* 2004, 4, 911–4.
313. D. Zhang, K. Ryu, X. Liu, E. Polikarpov, J. Ly, M. E. Tompson and C. Zhou, Transparent, conductive, and flexible carbon nanotube films and their application in organic light-emitting diodes, *Nano Lett.* 2006, 6, 1880–6.
314. A. D. Pasquier, H. E. Unalan, A. Kanwal, S. Miller and M. Chhowalla, Conducting and transparent single-wall carbon nanotube electrodes for polymerfullerene solar cells, *Appl. Phys. Lett.* 2005, 87, 203511.
315. M. W. Rowell, M. A. Topinka and M. D. McGehee, Organic solar cells with carbon nanotube network electrodes, *Appl. Phys. Lett.* 2006, 88, 233506.

316. Z. Yue, G. Wu, X. Chen, Y. Han, L. Liu and Q. Zhou, Facile, room-temperature synthesis of $NiSe^2$ nanoparticles and its improved performance with graphene in dye-sensitized solar cells, *Mater. Lett.* 2017, 192, 84–7.

317. M. Batmunkh, M. J. Biggs and J. G. Shapter, Carbon nanotubes for dye-sensitized solar cells, *Small* 2015, 11, 2963–89.

318. F. N. Xia, T. Mueller, Y. M. Lin, A. V.-Garcia and P. Avouris, Ultrafast, graphene photodetector, *Nat. Nanotechnol.* 2009, 4, 839–43.

319. T. Mueller, F. Xia and P. Avouris, Graphene photodetectors for high-speed optical communications, *Nat. Photonics* 2010, 4, 297–301.

320. Z. J. Liang, H. X. Liu, K. M. Liu, Y. X. Niu and Y. H. Yin, The analysis of microcavity-integrated graphene photodetector's SNR based on 1.06 μm, *Spectrosc. Spect. Anal.* 2017, 37, 356–60.

321. Z. F. Chen, X. M. Li, J. Q. Wang, L. Tao, M. Z. Long, S. J. Liang, L. K. Ang, C. Shu, H. K. Tsang and J. B. Xu, Synergistic effects of plasmonics and electron trapping in graphene short-wave infrared photodetectors with ultrahigh responsivity, *ACS Nano* 2017, 11, 430–7.

322. S. Liang, Z. Ma, G. Wu, N. Wei, L. Huang, H. Huang, H. Liu, S. Wang and L.-M. Peng, Microcavity-integrated carbon nanotube photodetectors, *ACS Nano* 2016, 10, 6963–71.

323. T. F. Zhang, Z. P. Li, J. Z. Wang, W. Y. Kong, G. A. Wu, Y. Z. Zheng, Y. W. Zhao, E. X. Yao, N. X. Zhuang and L. B. Luo, Broadband photodetector based on carbon nanotube thin film/single layer graphene Schottky junction, *Sci. Rep.* 2016, 6, 38569.

324. X. He and F. Leonard, Uncooled carbon nanotube photodetectors, *Adv. Opt. Mater.* 2015, 3, 989–1011.

325. X. Yu, Z. Dong, J. K. W. Yang and Q. J. Wang, Room-temperature mid-infrared photodetector in all-carbon graphene nanoribbon-C^{60} hybrid nanostructure, *Optica* 2016, 3, 979–84.

326. M. Liu, X. B. Yin, E. U.-Avila, B. S. Geng, T. Zentgraf, L. Ju, F. Wang and X. Zhang, A graphene-based broadband optical modulator, *Nature* 2011, 474, 64–7.

327. J. B. Liu, P. J. Li, Y. F. Chen, X. B. Song, Q. Mao, Y. Wu, F. Qi, B. Zheng, J. He, H. Yang, Q. Wen and W. Zhang, Flexible terahertz modulator based on coplanar-gate graphene field-effect transistor structure, *Opt. Lett.* 2016, 41, 816–9.

328. B. N. Szafranek, G. Fiori, D. Schall, D. Neumaier and H. Kurz, Current saturation and voltage gain in bilayer graphene field effect transistors, *Nano Lett.* 2012, 12, 1324–8.

329. F. Gianluca and I. Giuseppe, On the possibility of tunable-gap bilayer graphene FET, *IEEE Electron. Device Lett.* 2009, 30, 261–4.

330. C. Martina, F. Gianluca and I. Giuseppe, A semianalytical model of bilayerz-graphene field-effect transistor, *IEEE Electron. Device Lett.* 2009, 56, 2979–86.

331. F. Pasadasa and D. Jimenez, Large-signal model of the bilayer graphene field-effect transistor targeting radio-frequency applications: Theory versus experiment, *J. Appl. Phys.* 2016, 118, 244501.

332. M. A. Velasco-Soto, S. A. Perez-Garcia, J. Alvarez-Quintana, Y. Cao, L. Nyborg and L. Licea-Jimenez, Selective band gap manipulation of graphene oxide by its reduction with mild reagents, *Carbon* 2015, 93, 967–73.

333. N. T. T. Tran, S. Y. Lin, O. E. Glukhova and M. F. Lin, p-Bonding-dominated energy gaps in graphene oxide, *RSC Adv.* 2016, 6, 24458.

334. T. Q. Trung, N. T. Tien, D. Kim, M. Jang, O. J. Yoon and N.-E. Lee, A flexible reduced graphene oxide field-effect transistor for ultrasensitive strain sensing, *Adv. Funct. Mater.* 2014, 24, 117124.

335. T. K. Truong, T. Nguyen, T. Q. Trung, I. Y. Sohn, D. J. Kim, J. H. Jung and N.-E. Lee, Reduced graphene oxide field-effect transistor with indium tin oxide extended gate for proton sensing, *Curr. Appl. Phys.* 2014, 14, 738–43.

336. D. Joung, A. Chunder, L. Zhai and S. I. Khondaker, High yield fabrication of chemically reduced graphene oxide field effect transistors by dielectrophoresis, *Nanotechnology* 2010, 21, 165202.

337. W. Gao, N. Singh, L. Song, Z. Liu, A. L. M. Reddy, L. Ci, R. Vajtai, Q. Zhang, B. Wei and P. M. Ajayan, Direct laser writing of micro-supercapacitors on hydrated graphite oxide films. *Nat. Nanotechnol.* 2011, 6, 496–500.

338. Y. Chen, X. Zhang, D. Zhang, P. Yu and Y. Ma, High performance supercapacitors based on reduced graphene oxide in aqueous and ionic liquid electrolytes, *Carbon* 2011, 49, 573–80.

339. M. Veerapandian, M. H. Lee, K. Krishnamoorthy and K. Yun, Synthesis, characterization and electrochemical properties of functionalized graphene oxide, *Carbon* 2012, 50, 4228–8.

340. J. T. Robinson, F. K. Perkins, E. S. Snow, Z. Wei and P. E. Sheehan, Reduced graphene oxide molecular sensors, *Nano Lett.* 2008, 8, 3137–40.

341. I. P. Murray, S. J. Lou, L. J. Cote, S. Loser, C. J. Kadleck, T. Xu, J. M. Szarko, B. S. Rolczynski, J. E. Johns, J. Huang, L. Yu, L. X. Chen, T. J. Marks and M. C. Hersam, Graphene oxide interlayers for robust, high-efficiency organic photovoltaics, *J. Phys. Chem. Lett.* 2011, 2, 3006–12.

342. Z. Yin, S. Sun, T. Salim, S. Wu, X. Huang, Q. He, Y. M. Lam and H. Zhang, Organic photovoltaic devices using highly flexible reduced graphene oxide films as transparent electrodes, *ACS Nano* 2010, 4, 5263–8.

343. S. K. Saha, S. Bhaumik, T. Maji, T. K. Mandal and A. J. Pal, Solution-processed reduced graphene oxide in light-emitting diodes and photovoltaic devices with the same pair of active materials, *RSC Adv.* 2014, 4, 35493–9.

344. Y. Lu, Y. Jiang, W. Wei, H. Wu, M. Liu, L. Niu and W. Chen, Novel blue light emitting graphene oxide nanosheets fabricated by surface functionalization, *J. Mater. Chem.* 2012, 22, 2929–34.

345. M. C. Lin, M. Gong, B. Lu, Y. Wu, D. Y. Wang, M. Guan, M. Angell, C. Chen, J. Yang, B. J. Hwang and H. Dai, An ultrafast rechargeable aluminium-ion battery, *Nature* 2015, 520, 324–8.

346. J. V. Rani, V. Kanakaiah, T. Dadmal, M. S. Rao and S. Bhavanarushi, Fluorinated natural graphite cathode for rechargeable ionic liquid based aluminum-ion battery, *J. Electrochem. Soc.* 2013, 160, A1781–4.

347. Z. Ao and F. Peeters, High-capacity hydrogen storage in Al-adsorbed graphene, *Phys. Rev. B* 2010, 81, 205406.

348. Z. Ao, Q. Jiang, R. Zhang, T. Tan and S. Li, Al doped graphene: A promising material for hydrogen storage at room temperature, *J. Appl. Phys.* 2009, 105, 4307.

349. T. Dürkop, S. A. Getty, Enrique Cobas and M. S. Fuhrer, Extraordinary mobility in semiconducting carbon nanotubes, *Nano Lett.* 2004, 4, 35–9.

350. S. J. Tans, A. R. M. Verschueren and C, Dekker, Room-temperature transistor based on a single carbon nanotube, *Nature* 1998, 393, 49–52.

351. R. Martel, T. Schmidt, H. R. Shea, T. Hertel and P. Avouris, Single- and multi-wall carbon nanotube field-effect transistors, *Appl. Phys. Lett.* 1998, 73, 2447.

352. M. Shlafman, T. Tabachnik, O. Shtempluk, A. Razin, V. Kochetkov and Y. E. Yaish, Self aligned hysteresis free carbon nanotube field-effect transistors, *Appl. Phys. Lett.* 2016, 108, 163104.

353. M. Myodo, M. Inaba, K. Ohara, R. Kato, M. Kobayashi, Y. Hirano, K. Suzuki and H. Kawarada, Large-current-controllable carbon nanotube field-effect transistor in electrolyte solution, *Appl. Phys. Lett.* 2015, 106, 213503.

354. Y. Niimi, T. Matsui, H. Kambara, K. Tagami, M. Tsukada and H. Fukuyama, Scanning tunneling microscopy and spectroscopy of the electronic local density of states of graphite surfaces near monoatomic step edges, *Phys. Rev. B* 2006, 73, 085421.

355. J. A. Misewich, R. Martel, P. Avouris, J. C. Tsang, S. Heinze and J. Tersoff, Electrically induced optical emission from a carbon nanotube FET, *Science* 2003, 300, 783–6.

356. R. Yuksel, Z. Sarioba, A. Cirpan, P. Hiralal and H. E. Unalan, Transparent and flexible supercapacitors with single walled carbon nanotube thin film electrodes, *ACS Appl. Mater. Interfaces*, 2014, 6, 15434–9.

357. T. Mueller, M. Kinoshita, M. Steiner, V. Perebeinos, A. A. Bol, D. B. Farmer and P. Avouris, Efficient narrow-band light emission from a single carbon nanotube p-n diode, *Nat. Nanotechnol.* 2010, 5, 27–31.

358. F. Pyatkov, V. Futterling, S. Khasminskaya, B. S. Flavel, F. Hennrich, M. M. Kappes, R. Krupke and W. H. P. Pernice, Cavity-enhanced light emission from electrically driven carbon nanotubes, *Nat. Photonics* 2016, 10, 420–7.

359. S. Wang, Q. Zeng, L. Yang, Z. Zhang, Z. Wang, T. Pei, L. Ding, X. Liang, M. Gao, Y. Li and L. M. Peng, High-performance carbon nanotube light-emitting diodes with asymmetric contacts, *Nano Lett.* 2011, 11, 23–9.

360. J. Mittal and K. L. Lin, Carbon nanotube-based interconnections, *J. Mater. Sci.* 2017, 52, 643–62.

361. S. Karmjit and R. Balwinder, Temperature-dependent modeling and performance evaluation of multi-walled CNT and single-walled CNT as global interconnects, *J. Electron. Mater.* 2015, 44, 4825–35.

362. W. S. Hwang, P. Zhao, K. Tahy, L. O. Nyakiti, V. D. Wheeler, R. L. Myers-Ward, C. R. Eddy Jr., D. K. Gaskill, J. A. Robinson, W. Haensch, H. (Grace) Xing, A. Seabaugh and D. Jena, Graphene nanoribbon field-effect transistors on wafer-scale epitaxial graphene on SiC substrates, *APL Mater.* 2015, 3, 011101.

363. T. Shimizu, J. Haruyama, D. C. Marcano, D. V. Kosinkin, J. M. Tour, K. Hirose, et al., Large intrinsic energy bandgaps in annealed nanotube-derived graphene nanoribbons, *Nat. Nanotechnol.* 2011, 6, 45–50.

364. X. R. Wang, Y. J. Ouyang, X. L. Li, H. L. Wang, J. Guo and H. J. Dai, Room-temperature all-semiconducting sub-10-nm graphene nanoribbon field-effect transistors, *Phys. Rev. Lett.* 2008, 100, 206803.

365. M. Y. Han, B. Özyilmaz, Y. Zhang and P. Kim, Energy band-gap engineering of graphene nanoribbons, *Phys. Rev. Lett.* 2007, 98, 206805.

366. C. Berger, Z. Song, X. Li, X. Wu, N. Brown, C. Naud, D. Mayou, T. Li, J. Hass, A. N. Marchenkov, E. H. Conrad, P. N. First and W. A. de Heer, Electronic confinement and coherence in patterned epitaxial graphene, *Science* 2006, 312, 1191–6.

367. C. Gao, L. Li, A. R. O. Raji, A. Kovalchuk, Z. Peng, H. Fei, Y. He, N. D. Kim, Q. Zhong, E. Xie and J. M. Tour, Tin disulfide nanoplates on graphene nanoribbons for full lithium ion batteries, *ACS Appl. Mater. Interfaces* 2015, 7, 26549–56.

368. Y. Yang, L. Li, H. L. Fei, Z. W. Peng, G. D. Ruan and J. M. Tour, Graphene nanoribbon/V^2O^5 cathodes in lithium-ion batteries, *ACS Appl. Mater. Interfaces* 2014, 6, 9590–4.

369. L. Li, A. R. O. Raji and J. M. Tour, Graphene-wrapped MnO^2-graphene nanoribbons as anode materials for high-performance lithium ion batteries, *Adv. Mater.* 2013, 25, 6298–302.

370. J. Y. Wu, S. C. Chen, O. Roslyak, G. Gumbs and M. F. Lin, Plasma excitations in graphene: Their spectral intensity and temperature dependence in magnetic field, *ACS Nano* 2011, 5, 1026–32.

371. J. H. Ho, C. P. Chang and M. F. Lin, Electronic excitations of the multilayered graphite, *Phys. Lett. A* 2006, 352, 446–50.

372. T. N. Do, C. P. Chang, P. H. Shih and M. F. Lin, Stacking-enriched magneto-transport properties of few-layer graphenes, arXiv:1704.01313.

373. J. Y. Wu, S. C. Chen, O. Roslyak, G. Gumbs and M. F. Lin, Plasma excitations in graphene: Their spectral intensity and temperature dependence in magnetic field, *ACS Nano* 2011, 5, 1026–32.

374. C. Zheng, X. F. Zhou, H. L. Cao, G. H. Wang and Z. P. Liu, Edge-enriched porous graphene nanoribbons for high energy density supercapacitors, *J. Mater. Chem. A* 2014, 2, 7484–90.

375. P. Ahuja, R. K. Sharma and G. Singh, Solid-state, high-performance supercapacitor using graphene nanoribbons embedded with zinc manganite, *J. Mater. Chem. A* 2015, 3, 4931–7.

376. P. Pachfule, D. Shinde, M. Majumder and Q. Xu, Fabrication of carbon nanorods and graphene nanoribbons from a metal-organic framework, *Nat. Chem.* 2016, 8, 718–24.

377. L. Tao, E. Cinquanta, D. Chiappe, C. Grazianetti, M. Fanciulli, M. Dubey, A. Molle and D. Akinwande, Silicene field-effect transistors operating at room temperature. *Nat. Nanotechnol.* 2015, 10, 227–31.

378. L. F. Li, S. Z. Lu, J. B. Pan, Z. H. Qin, Y. Q. Wang, Y. L. Wang, G. Y. Cao, S. X. Du and H. J. Gao, Buckled germanene formation on Pt(111), *Adv. Mater.* 2014, 26, 4820–8.

379. P. Vogt, P. D. Padova, C. Quaresima, J. Avila, E. Frantzeskakis, M. C. Asensio, A. Resta, B. Ealet and G. Le Lay, Silicene: Compelling experimental evidence for graphenelike two-dimensional silicon, *Phys. Rev. Lett.* 2012, 108, 155501.

380. F. Zhu, W. Chen, Y. Xu, C. Gao, D. Guan, C. Liu, D. Qian, S. Zhang and J. Jia, Epitaxial growth of two-dimensional stanine, *Nat. Mater.* 2015, 14, 1020–5.

381. M. Agostini, M. Allardt, E. Andreotti, A. M. Bakalyarov, M. Balata and I. Barabanov, Results on neutrinoless double-β decay of ^{76}Ge from phase I of the GERDA experiment, *Phys. Rev. Lett.* 2013, 111, 057005.

382. J. Y. Wu, G. Gumbs and M. F. Lin, Combined effect of stacking and magnetic field on plasmon excitations in bilayer graphene, *Phys. Rev. B* 2014, 89, 165407.

383. M. F. Lin, Y. C. Chuang and J. Y. Wu, Electrically tunable plasma excitations in AA-stacked multilayer graphene, *Phys. Rev. B* 2012, 86, 125434.

384. Y. C. Chuang, J. Y. Wu and M. F. Lin, Electric field dependence of excitation spectra in AB-stacked bilayer graphene. *Sci. Rep.* 2013, 3, 1368.
385. Y. Xu, B. Yan, H. J. Zhang, J. Wang, G. Xu, P. Tang, W. Duan and S. C. Zhang, Large-gap quantum spin Hall insulators in tin films, *Phys. Rev. Lett.* 2013, 111, 136804.
386. H. Wang, L. Yu, Y. H. Lee, Y. Shi, A. Hsu, M. L. Chin, L. J. Li, M. Dubey, J. Kong and T. Palacios, Integrated circuits based on Bi-layer MoS_2 transistors, *Nano Lett.* 2012, 12, 4674–80.

Index

π-electronic absorption peaks, 34

1D energy subbands, 97, 99
1D Landau-subbands, 9, 62, 114
2D AB-stacked graphenes, 56
2D graphenes, 114
2D multilayer AA-stacked graphene, 32
$2N_y$ tight-binding functions, 96
3D Dirac cones, 70, 107
3D energy dispersions, 107

AA-stacked graphenes, 36, 50, 113, 117
AA-stacked graphite, 2–3, 29, 33–34, 38,
        43, 107–109, 113, 115, 117
ABC-stacked graphene, 36, 73, 74, 81,
        109, 117
ABC-stacked graphite, 4–5, 107, 115, 117
Absorption spectra, 34
Absorption spectroscopy, 25–27
AB-stacked graphene, 36, 50, 65, 83, 109
AB-stacked graphite, 2, 3, 33, 107,
        108, 115
Aharonov–Bohm effect, 89
Angle-resolved photoemission
        spectroscopy (ARPES), 24–25,
        33, 117
Anisotropic Dirac cone, 71–73
Armchair graphene nanoribbons, 96, 99,
        103–104, 117
ARPES, see Angle-resolved
        photoemission spectroscopy
        (ARPES)

Beating oscillations, 43–44
Bernal graphite, 3, 13–15
    electronic structures without
        external fields, 51–54
    magnetic quantization, 57–62
        anti-crossings of Landau
            subbands, 60–62
        Landau subbands and wave
            functions, 57–60

magneto-optical properties, 62–67
optical properties without external
        fields, 54–57
overview, 51
Binnig, G., 22

Carbon nanotubes, 5
    magneto-electronic properties
        of, 85–91
    magneto-optical spectra of, 91–96

Density of states (DOS), 3, 6, 24, 29,
        33–34, 41, 53, 59, 74–75, 91,
        99, 107
Dimensional crossover, 73–75
Dirac-cone structure, 2, 5, 29, 32, 33,
        35–36, 72–73, 75–76, 78–80, 108
Dirac fermions, 62, 67
Dirac quasiparticles, 56, 58
DOS, see Density of states (DOS)

Electric polarization, 93
Electron-hole asymmetry, 67
Electronic structures, without
        external fields
    Bernal graphite, 51–54
    rhombohedral graphite, 69–70
    simple hexagonal graphite, 29–34
Energy dispersions, 88
Energy gap, 88–89

Fermi's golden rule, 26
Fermi velocity, 84
Field-effect transistor (FET), 110–111
Fourier-transform spectrometer, 26

Gradient approximation, for optical
        properties, 18–22
Graphene nanoribbons, 5–6
    magneto-electronic properties of,
        96–99
    magneto-optical spectra of, 99–107

Graphite, 36, 65
  composites, 110
  crystals, 2
  fibers, 118
Graphite intercalation compounds,
    110, 118

Hamiltonian matrix, 12–13, 86

Infrared spectroscopy, 56
Interband absorption, 43, 46

J-decoupled standing waves, 94
Joint density of states (JDOS), 76

Kubo formula, 18

Landau levels (LLs), and wave
    functions, 4, 5, 6, 36–38, 50,
    102, 114
Landau-subbands (LSs), 4, 6, 9, 29, 58, 61,
    67, 107–108
  anti-crossings of, 60–62
  energy spectra, 38–43, 76, 78
  and wave functions, 57–60
Layered graphites, magnetic tight-
    binding model for, 9–22
  Bernal graphite, 13–15
  gradient approximation for optical
    properties, 18–22
  rhombohedral graphite, 15–17
  simple hexagonal graphite, 12–13
LLs, *see* Landau levels (LLs)
LSs, *see* Landau-subbands (LSs)

Magnetic quantization
  Bernal graphite
    anti-crossings of Landau
      subbands, 60–62
    Landau subbands and wave
      functions, 57–60
  simple hexagonal graphite
    Landau levels and wave
      functions, 36–38
    Landau-subband energy spectra,
      38–43
Magnetic tight-binding model, for
    layered graphites, 9–22
  Bernal graphite, 13–15

gradient approximation for optical
    properties, 18–22
rhombohedral graphite, 15–17
simple hexagonal graphite, 12–13
Magneto-absorption spectroscopy, 26
Magneto-electronic properties, 76–82
  Onsager quantization, 78–82
  tight-binding model, 76–78
Magneto-optical properties
  Bernal graphite, 62–67
  rhombohedral graphite, 82–84
  simple hexagonal graphite, 43–50
Magneto-Raman spectroscopy, 27
Monolayer graphene, 1, 29, 34, 38

Natural graphite, 2
N-layer ABC-stacked graphene, 76
N-layer AB-stacked graphenes, 53

Onsager quantization, 6, 78–82
Optical properties
  gradient approximation for, 18–22
  without external fields
    Bernal graphite, 54–57
    rhombohedral graphite, 75–76
    simple hexagonal graphite, 34–36

Photodiode array spectrophotometer, 26
Photoelectrons, 24
Pierce, D. T., 22
p-polarized excitation beam, 55
Pristine graphite, 110, 118
Pseudo-spin polarizations, 78

QHE, *see* Quantum Hall effect (QHE)
QLLs, *see* Quasi-Landau levels (QLLs)
Quantum confinement, 85–112
  in carbon nanotubes, 85–96
    magneto-electronic properties,
      85–91
    magneto-optical spectra, 91–96
  comparisons and applications,
    107–112
  in graphene nanoribbons, 96–107
    magneto-electronic properties,
      96–99
    magneto-optical spectra, 99–107
  overview, 85
Quantum Hall effect (QHE), 119

Quantum tunneling current, 22
Quasi-Landau levels (QLLs), 85, 97, 99,
    102, 107, 110, 116

Rhombohedral graphite, 15–17
    anisotropic Dirac cone, 71–73
    dimensional crossover, 73–75
    electronic structures without
        external fields, 69–70
    magneto-electronic properties, 76–82
        Onsager quantization, 78–82
        tight-binding model, 76–78
    magneto-optical properties, 82–84
    optical properties without external
        fields, 75–76
    overview, 69
Rohrer, H., 22

Scanning tunneling microscopy (STM),
    4, 22–24
Scanning tunneling spectroscopy (STS),
    22–24, 41, 90, 117
Simple hexagonal graphite, 2, 12–13
    electronic structures without
        external fields, 29–34
    magnetic quantization, 36
        Landau levels and wave
            functions, 36–38
        Landau-subband energy spectra,
            38–43

magneto-optical properties, 43–50
optical properties without external
    fields, 34–36
overview, 29
Slonczewski–Weiss–McClure (SWM)
    Hamiltonian, 4
Spin-dependent spectroscopic mode, 22
s-polarized photoresponse, 55
State degeneracy, 88
STM, *see* Scanning tunneling
    microscopy (STM)
STS, *see* Scanning tunneling
    spectroscopy (STS)
Subenvelope functions, 102
SWM, *see* Slonczewski–Weiss–McClure
    (SWM) Hamiltonian
SWM model, 14, 64, 67, 115

Threshold absorption, 94
Tight-binding model, 1, 2, 4, 6, 76–78

van Hove singularities (VHSs), 91
Velocity matrix element, 34
VHSs, *see* van Hove singularities (VHSs)

Wiesendanger, R., 23

Zigzag graphene nanoribbons, 96, 97, 99,
    103–104, 106, 117